THE U.S. SUBMARINE PRODUCTION BASE

An Analysis of Cost, Schedule, and Risk for Selected Force Structures

T0308284

John Birkler, John Schank, Giles Smith,
Fred Timson, James Chiesa, Marc
Goldberg, Michael Mattock,
Malcolm MacKinnon

Prepared for the
Office of the Secretary of Defense

National Defense Research Institute

RAND Approved for public release; distribution unlimited

The research described in this report was sponsored by the Office of the Secretary of Defense under RAND's National Defense Research Institute, a federally funded research and development center supported by the Office of the Secretary of Defense, the Joint Staff, and the defense agencies, Contract MDA903-90-C-0004.

ISBN: 0-8330-1548-6

RAND is a nonprofit institution that helps improve policy and decisionmaking through research and analysis. RAND® is a registered trademark. RAND's publications do not necessarily reflect the opinions or policies of its research sponsors.

Cover design by Corrine Maier

Published 1994 by RAND
1700 Main Street, P.O. Box 2138, Santa Monica, CA 90407-2138
1200 South Hayes Street, Arlington, VA 22202-5050
201 North Craig Street, Suite 202, Pittsburgh, PA 15213-1516
RAND URL: http://www.rand.org/
To order RAND documents or to obtain additional information, contact
Distribution Services: Telephone: (310) 451-7002; Fax: (310) 451-6915; Email:
order@rand.org

In January 1993, RAND's National Defense Research Institute (NDRI) was asked by the Office of the Under Secretary of Defense for Acquisition (now Acquisition and Technology) to compare the practicality and cost of two approaches to future submarine production: (1) allowing production to shut down as currently programmed submarines are finished, then restarting production when more submarines are needed, and (2) continuing low-rate production. The research was motivated by concerns that the submarine production base might not be easily reconstituted if production is shut down and by the countervailing recognition that deferring new submarine starts might yield substantial savings, particularly over the short term.

This report is a comprehensive record of the methods employed in RAND's analyses and the results obtained. Those analyses were completed and briefed to the research sponsors and other interested parties in the summer of 1993. They reflect what was known then about cost, schedules, and so forth. (The text of this report includes some information that has come to our attention since then.)

It is our intention that this report be understandable by someone with little knowledge of submarine production or cost and schedule analysis but that it satisfy those interested in the details of the assessments underlying the conclusions presented. It will be supplemented by two less comprehensive products—a shorter report that emphasizes results and takes a more selective approach to the supporting material offered and a "research brief," a single-sheet stand-alone summary of findings.

This research was carried out in NDRI's Acquisition and Technology Policy Center. The National Defense Research Institute is a federally funded research and development center sponsored by the Office of the Secretary of Defense, the Joint Staff, and the defense agencies.

CONTENTS

FIGURES

The current attack submarine production program is coming to an end. After decades of building three or more submarines annually, there have been no construction starts since 1991. It is generally believed that the current fleet of Los Angeles–class attack submarines is big enough to meet U.S. security needs for many years. Superficially, it may seem appropriate, especially given budgetary constraints, to suspend submarine production for a period of time.

At some point in the future, however, it will be necessary to build more submarines to replace the Los Angeles–class ships as they age and can no longer be operated with high standards of safety and reliability. Initiating such a construction program from scratch will involve serious challenges. Nuclear submarines are among the most complex structures built by man. Not only must they survive and function under water for long periods of time in a hostile environment, they contain a nuclear reactor in immediate proximity to the crew. Despite these challenges, U.S. nuclear submarines have demonstrated their reliability in diverse conflict situations while maintaining a very good safety record over the years. That history can be credited in large part to the highly skilled submarine design, engineering, and construction workforce, both in the shipyards and at the factories of critical-component vendors.

The most recently started submarine is now three years into construction. Shipyard workers and component vendors needed only in the initial phase of construction are already dispersing or preparing to exit the business. More will leave as the industry shuts down in phases. If more submarines are not started soon, then rebuilding the workforce, reopening the shipyard facilities, and reestablishing the vendor base could be very costly and time-consuming. Reconstitution could also compromise the reliability and safety of submarines constructed before today's high standards are reattained.

The purpose of this study was to determine the practicality of extending the current gap between submarine starts, given the time required to restart; estimate the money likely to be saved, given the offsetting costs of shutdown and restart; and characterize the largely unquantifiable risks involved in a reconstitution strategy. Our conclusions are as follows:

- It takes so long to restart production after shutdown that construction of the next class of submarines must be started by around 2001 if fleet sizes that the government judges consistent with anticipated national security needs are to be sustained.

- For the longest gaps feasible, the discounted stream of costs required to sustain the submarine force to 2030 results in savings of less than a billion dollars compared to the cost of a more continuous program. That is well within the margin of error with which we can now project such costs.

- Given the difficulties and challenges involved in restarting submarine production from scratch, our cost estimates for restart may be too low and our schedule estimates too optimistic. Further risks related to nuclear licensing and environmental and safety concerns may jeopardize the success of the nuclear submarine program.

- Considering the limited savings realizable and the substantial risks incurred in extended-gap scenarios, we recommend that construction of additional submarines be started soon. Specifically, we recommend that the third Seawolf-class submarine, now planned for a 1996 start, be funded, and that the Navy proceed with plans for beginning a new class of submarines in the late 1990s.

In arriving at these conclusions, we drew on quantitative data and qualitative information from private- and public-sector shipyards and vendors, relevant components of the U.S. Navy and the Office of the Secretary of Defense, and foreign governments with shutdown experience. Sources included persons with varying perspectives on the seriousness of the delays, costs, and risks associated with a production gap. We critically reviewed all data, made adjustments as appropriate, and built and ran models to draw inferences where the nature of the data permitted it. We determined how stopping and restarting production affects shipyard and vendor costs and schedules and how decisions about future fleet size and production rate affect the production gaps feasible. These results were then combined to yield discounted cost streams for sustaining the submarine production base under a strategy of continued production and under various gapping strategies. We accounted for the costs of producing, operating, and maintaining the submarine force until 2030, when the Los Angeles–class submarines will all have been retired. The results of the analyses underlying our principal conclusions are as follows.

Shipyard Effects. If submarine production is to be suspended for a number of years, substantial sums will have to be expended to shut down shipyard activities and facilities in a manner that preserves tooling and information to facilitate restart. Then, the yard and its production lines will have to be maintained in working order during the gap. Additional expenses will be incurred in reopening facilities and rebuilding the workforce at the end of the gap. These workforce expenses dominate the total (for an illustrative case, see Figure S.1). Costs of rebuilding the workforce include those of hiring and training new workers, plus those arising from inefficiencies in producing early submarines (when the workforce will have more workers at lower levels of productivity than it will later). We found that submarine production restart costs can be reduced if shipyards remain active with aircraft carrier construction or with submarine overhauls. (Currently, the latter are performed in Navy shipyards.)

The longer the production gap, the more skilled workers will be permanently lost from the industrial base, and the longer it will take to produce the first submarine and to ramp production up to the desired rate. If workers can be retained through other shipyard activities, these delays can be reduced. For example, whereas it would take over ten years after contract award to deliver the first submarine starting from a residual skilled workforce of 250, it would take only six years if 1000 skilled workers could be retained.

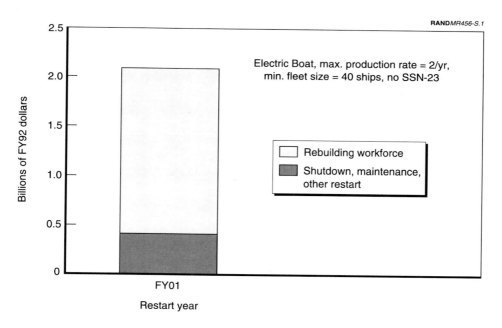

Figure S.1—Total Cost to Shut Down, Maintain, and Restart a Shipyard

Vendor Effects. Shipyards buy or receive through the government many submarine components—nuclear and nonnuclear—produced by outside suppliers. To be ready for installation at the correct point in submarine construction, work on some key nuclear components must begin well in advance (see Figure S.2). Current work will keep nuclear-system vendors busy for the next two or three years (assuming a new aircraft carrier is built). Design work has already begun on the longest-lead components for a new attack submarine. Unless there is a lengthy production gap, it would not be practical to shut down the suppliers of such components. Neither is it necessary to shut down the sole remaining U.S. producer of naval reactor cores, as that firm is engaged in producing cores to refuel aircraft carriers and the Trident-missile-carrying submarines. Shutting the remaining nuclear vendors down for several years would result in hundreds of millions of dollars in reconstitution costs, assuming reconstitution is feasible at all.

Although the nuclear-vendor base is small, there are on the order of a thousand suppliers of nonnuclear submarine-specific components. For the most part,

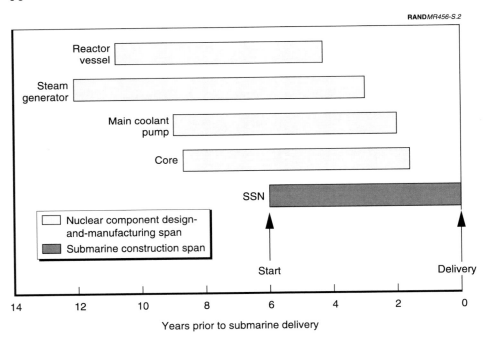

NOTE: The times assume an active industrial base; required lead times could be longer following an extended production gap.

Figure S.2—SSN-21 Shipyard Need Dates and Design-and-Manufacturing Spans for Selected Nuclear Components

supply of these latter components could be quickly resumed once demand for them is renewed following a production gap. A few, however, require special skills or technologies that may be difficult to recover should the firms producing them go out of business during a gap. For these cases, comprising at least a few products and at most a few dozen, reconstitution costs could amount to half a billion dollars.

If submarine orders are delayed, the government could take a variety of actions that could help avoid the need to reconstitute the nuclear and nonnunclear vendor bases. Such measures include funding the production of items in advance of need, paying the firms to develop and prototype advanced methods to manufacture the needed components, or allocating other Navy work to those firms. Each of these measures has its drawbacks. But whatever is chosen, it must be done soon, as critical nonnuclear suppliers may otherwise begin to go out of business within the next year.

Effects of Fleet Size and Production Rate on Delivery Gap. We have referred to the "production gap" that began in 1991 and will extend until construction on the next submarine starts. Since fleet size effects are determined by the time of submarine entry into the force, we now refer to the "delivery gap," or time between delivery of the last submarine now under construction and the next one.

Fleet size, maximum sustained production rate, and delivery gap are interrelated. The implications for gap length cannot be understood without understanding the constraints that production rate places on fleet size. Estimates of future required attack submarine fleet size range roughly from 40 to 60. Given the rate at which submarines will be retired in the future, a production rate of one submarine per year following a 1998 restart cannot sustain a fleet size of 30 (see Figure S.3).[1] Two per year will sustain 40 but not 50; it takes three per year to sustain 60. If the service lives of the more recently built submarines could be extended from a maximum of 30 years to 35 years, the fleet size sustainable at a given production rate would increase. A fleet size of 50, for example, could then be sustained at two new submarines per year. However, *extending the lives of nuclear submarines is not a trivial task.* Much additional technical study and analysis of cost and military effectiveness is required before a decision could be made to implement it.

---

[1]In steady state, one new submarine per year could sustain a fleet of 30 submarines with 30-year lives. However, submarines of the Los Angeles class were built at an average rate of three per year and will be decommissioned at least as rapidly. At a production rate of one per year and a retirement rate of three per year, the fleet will shrink until all current ships are decommissioned (in 2027).

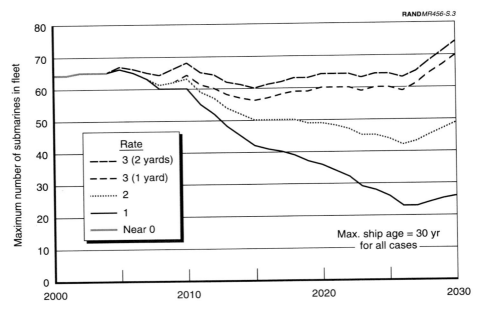

Figure S.3—Production Rate Influences the Fleet Size That Can Be Sustained

Figure S.4 shows the latest possible delivery date for the next submarine if various fleet sizes are to be maintained at a maximum production rate of two or three ships per year from a single shipyard, with a maximum ship life of 30 or 35 years. For several combinations of production rate, fleet size, and service life, it is not possible to sustain the fleet size minimum unless the first new attack submarine is delivered before 2004, which is impractical. (Such combinations are represented by the blank triangles in Figure S.4.) Maximum gaps are to 2010 if a 40-sub fleet is to be sustained and to 2007 if a 50-sub fleet is the objective. Given the inefficiencies of restart, such gaps mean that *construction* of the next submarine must start by 2001 at the latest.

For each of the maximum gaps shown in Figure S.4, it is possible to define a corresponding minimum gap as a baseline against which the savings of an extended gap can be compared. For example, as the figure shows, if an eventual fleet size of 40 is to be sustained at a maximum production rate of two ships per year, the delivery gaps must end in 2005 or 2010. The earlier date holds if submarine life is to be held to 30 years and the later if the more recently built subs can be extended to 35 years. The gap from delivery of the last ship currently under construction, scheduled for 1998, is then 7 years in the 30-year case and 12 years in the 35-year case (for the latter, see the lower bar in Figure S.5). The minimum gap achievable in either case entails initiating construction of a Seawolf-class submarine in 1996. The Seawolf's delivery date of 2002 would

RAND*MR456-S.4*

Latest year to deliver next submarine

For max. age of	30 yr / 35 yr	Production Rate	
		2 per yr	3 per yr
To sustain a fleet size of	40	2005 / 2010	2010 / 2010
	50	2005	2006 / 2007
	60		2004

NOTE: No third Seawolf; blank triangle indicates delivery needed earlier than is feasible.

Figure S.4—No Matter the Scenario, Restart Cannot Be Long Delayed

then result in a four-year delivery gap, followed by a three-year gap (upper bars in Figure S.5).

Gap Savings. Assuming the current submarine service life, a 40-ship fleet, and a two-per-year production rate, the maximum gap strategy saves about $700 million (net present value [NPV]) relative to the minimum gap case; for the 35-year option, roughly $200 million (see Figure S.6). These savings take into account all costs related to production restart, construction, and fleet operations and maintenance through 2030. The savings for both cases are much smaller than the uncertainty to which our projections are subject and the $2 billion savings achievable through extending ship life by five years.

When a production rate of three ships per year is allowed, extending the gap does not always result in savings, but the difference is, in all cases we examined, less than a billion dollars. Life extension, on the other hand, results in savings ranging from about a billion to about two and a half billion dollars, depending on the case.

Gap Risks and Constraints. The modest savings from extending the production and delivery gaps are achieved at a substantial increase in program risk. Some of this risk arises from the inherent uncertainty in making any kind of cost or

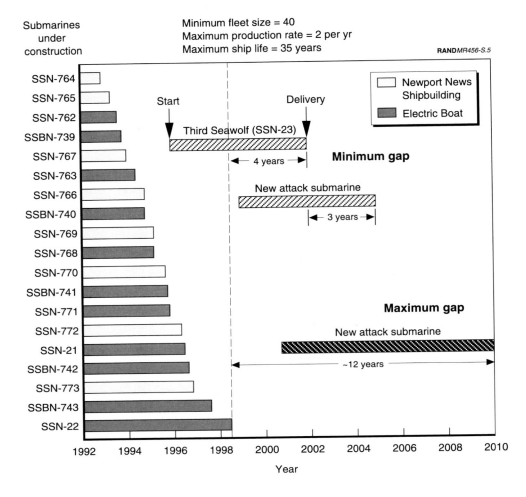

**Figure S.5—Gapping and Restart Relations Between Minimum-
and Maximum-Gap Strategies**

schedule estimate for an action that has no real analogue: No dormant industries have experienced production restarts recently. Also, we have made no allowance for problem resolution in our estimates, although British experience
indicates that it would be challenging to produce submarines that integrate
new technologies developed during the gap years.

Other risks relate to more specific infrastructure failures that could substantially postpone or even jeopardize a restart program's successful completion.
For some of the longer gap scenarios, for example, submarine design and development skills may atrophy, further lengthening the production phase. It is
uncertain whether construction management, technical, and trade skills can be

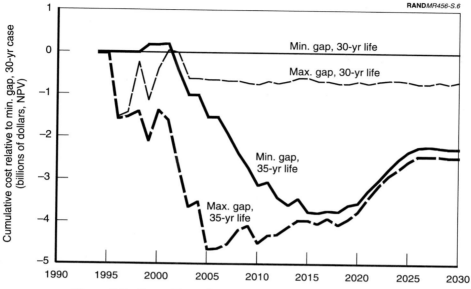

**Figure S.6—Long-Term Cost Differences Between Minimum
and Maximum Gaps Are Small**

reconstituted at any reasonable price; once firms and individuals leave the industry, it may not be possible to lure them back. Shipyard nuclear licenses and environmental permits may be lost if production is suspended; considering the urban locations of the shipyards, restoring those permits could be characterized conservatively as a serious political challenge. If restarting production at a lower skill level results in an eventual accident, particularly one involving a nuclear reactor, the ship's crew and everyone else in the vicinity could be endangered, and public pressure could halt submarine construction and curtail operations indefinitely.

Gapping production also constrains the fleet sizes and production rates that can be chosen. World events may lead to a decision that a fleet size of 60 is needed to ensure national security. Such a fleet size cannot be sustained if construction on the next submarine is not initiated before 2000. Even for a 50-ship fleet, delaying the next submarine start to 2000 or beyond would require a production rate greater than two per year, and the same would be true of a 40-ship fleet if the current 30-year lifespan is retained. It is uncertain whether submarine production at three per year would be viewed as affordable, and such a program would produce a full fleet of 30-year-lifespan submarines in less than 20 years, resulting in another production gap in the 2030s.

Recommendations. Given the limited savings achievable through gapping production and the substantial risks incurred, we recommend a "minimum gap" strategy that entails constructing the next Seawolf-class submarine beginning in 1996, to be followed by the first attack submarine incorporating a new design beginning around 1998. We also recommend that the Navy examine carefully the feasibility of extending the life of the more recently built attack submarines.

This work could not have been undertaken without the special relationship that exists between the Office of the Secretary of Defense (OSD) and RAND under the National Defense Research Institute (NDRI). For that relationship we are grateful. Many individuals in OSD and RAND deserve credit for the work discussed in this report. Their names and contributions would fill several pages. If we were to single out a senior person in OSD and another at RAND who participated in and supported this work in extraordinary ways, we would mention Gene Porter, Director, Acquisition Program Integration, and David Gompert, Director, NDRI.

We also want to thank the leadership and staff of the Office of the Secretary of Defense, the Office of the Secretary of the Navy, the Naval Sea Systems Command, the Navy Nuclear Propulsion Directorate, the Navy Program Executive Officer for Submarines, Electric Boat Division of General Dynamics, Newport News Shipbuilding, Mare Island Naval Shipyard, and Norfolk Naval Shipyard. The shipyards arranged for us to visit their facilities and gave us the opportunity to discuss production issues with those most directly involved. The shipyards and the Navy offices provided all the data we requested in a timely manner. We appreciate their sharing their perspectives with us and their treating different perspectives in a professional manner.

We are also indebted to the British and French Ministries of Defense for allowing us to visit their headquarters and submarine production facilities and to discuss their experiences with production gaps, low-rate production, and production issues.

This broad-based participation made possible the analysis described here.

Finally, we wish to thank RAND colleagues Joseph Large and James Winnefeld. Their thoughtful reviews occasioned many changes that improved the clarity of the report.

Chapter One

INTRODUCTION

The security and economic well-being of the United States depend upon freedom of the seas for merchant vessels engaged in American trade and U.S. warships defending American interests around the globe. The U.S. forces beneath
the world's oceans play a vital role in maintaining the American maritime pre-
eminence necessary to guarantee freedom of the seas. However, in light of
changes in the world, the accompanying reductions in threats to American interests and resources devoted to national defense, and the vigorous pace of
submarine construction in the past decade, there is no longer a pressing need
for production now of a new class of submarines for the U.S. fleet.[1]

—Donald J. Atwood, Jr.

Writing in 1992, then–Deputy Secretary of Defense Atwood expressed well the
need for U.S. attack submarines and the sufficiency of the current submarine
force to meet that need. At some point in the future, of course, it will be necessary to build more submarines to replace the current ones as they become too
old to operate safely. But how difficult will it be to resume production once it
has stopped? Important construction skills may be lost—skills that may be expensive and time-consuming to restore. Will the extra cost be greater than that
saved by stopping production? Would it be wiser to maintain skills and facilities by continuing to build submarines at a low rate in the interim, even though
building additional ships cannot be justified by near-term national security
needs?

These and related issues were addressed in the study reported in this volume,
which was undertaken by RAND's National Defense Research Institute for the
Office of the Under Secretary of Defense for Acquisition (now Acquisition and
Technology). RAND was asked to evaluate "the practicality and cost effective-

---

[1]Memorandum to Secretary of the Navy and others on "Submarine Forces for the Future," January
22, 1992.

ness of reconstitution of the submarine production base versus a continuing program for limited production." RAND's analysis built on earlier research on shutting down and restarting production in the aircraft industry[2] and on a broad set of studies in acquisition policy.

This report begins with some background information for the reader unfamiliar with the status of the U.S. submarine fleet and the special requirements of submarine construction. We then assess the cost and schedule implications of shutting down and restarting each of the principal elements of the submarine industrial base—the shipyards, nuclear-system vendors, and nonnuclear-system vendors. Next, we show how the schedule effects of shutdown and restart interact with factors such as desired fleet size, annual production rate, and number of operating shipyards to determine the maximum feasible gap (length of time) in submarine delivery. This analysis permits the construction of different time gap scenarios. We then estimate the costs of those scenarios, including the shutdown and restart costs calculated earlier. The quantitative analysis is combined with a qualitative assessment of important risks entailed in shutdown and restart to yield the report's conclusions and final observations.

Our analyses drew on information from shipyards and vendors, components of the U.S. Navy and the Office of the Secretary of Defense, and foreign governments with shutdown experience. Sources included persons with varying perspectives on the seriousness of the delays, costs, and risks associated with a production gap. We reviewed all data critically, made adjustments where we believed it appropriate, and built and ran models to draw inferences when the nature of the data permitted. Models and other methods specific to the individual analyses are discussed in connection with those analyses in the following chapters.

Although we consider a wide variety of factors, the focus is on the production base. We do discuss some of the implications of shutdown and restart for the nation's ability to design submarines (see, in particular, Appendix A), but not nearly in as much detail as we treat production. Clearly, design is required for production, and deterioration of the design base is as critical as losses to the production base.

We assume in our cost and schedule analyses that the motivation for restarting or continuing production is to replace obsolescent ships—or to preserve the capacity to replace them. We do not attempt to analyze the implications of a threat to national security dramatic enough to warrant some sort of crash program to bolster the attack submarine force. While the events of the past few

---

[2]John Birkler, Joseph Large, Giles Smith, and Fred Timson, *Reconstituting a Production Capability: Past Experience, Restart Criteria, and Suggested Policies*, RAND, MR-273-ACQ, 1993.

years indicate the volatility of the geostrategic environment, we believe that the emergence of such a threat in the near future is unlikely and, in any event, its implications are unclear. Suffice it to say that a national emergency could provide the impetus to build the submarine force up more quickly than we indicate, given sufficient resources.

The study's principal outputs were the analyses and conclusions specific to submarines, but we believe the approach we took is another important product. The overall conceptual framework and the individual models of workforce buildup and fleet dynamics could be used in analogous studies of other types of ships facing similar acquisition decisions. With appropriate modification for different production quantities, maintenance policies, and so forth, the analytical tools developed here could also be applied to other types of major weapon systems.

Chapter Two

BACKGROUND

This chapter provides some general background that may be helpful in understanding the analyses that follow. After reviewing the history and current status of U.S. submarine production and fleet composition, we discuss various pertinent aspects of submarine design, production, and operational life. Finally, we describe the evolution and organization of the submarine production base.

U.S. SUBMARINE PRODUCTION TO DATE[1]

The nuclear submarine propulsion system emerged in the early 1950s as the successor to the then-prevalent diesel-electric system, which used diesel engines on the surface and electric batteries while submerged. Nuclear power permitted a technical solution to the submarine's greatest vulnerabilities—the need to surface or snorkel periodically to recharge the batteries, and the submerged-speed constraint enforced by limited battery capacity. When the first nuclear submarine, USS *Nautilus*, was commissioned, the U.S. submarine fleet consisted of about 140 diesel-electric boats. New submarine classes were rapidly prepared for construction (see Figure 2.1). The Skate and Skipjack classes[2] were begun in the 1950s as refinements of the Nautilus concept. In 1958, construction started on the Thresher (now Permit) class, which was the first of what would today be considered the modern, front-line submarine. As the cold war with the Soviet Union raged, the development of the ballistic missile submarine became a national priority. Five classes of ballistic-missile-carrying submarines (SSBNs) were fielded, representing step improvements in propulsion technology and ship and missile design. In all, 41 SSBNs were commissioned between 1960 and 1967. During this same period, 24 nuclear attack submarines (SSNs) were commissioned to augment the eight nuclear-powered submarines built before 1960.

---

[1]For a historical overview of submarine missions and construction funding, see Appendix B.

[2]A submarine's class is denoted by the name (or number) of the first ship of that type.

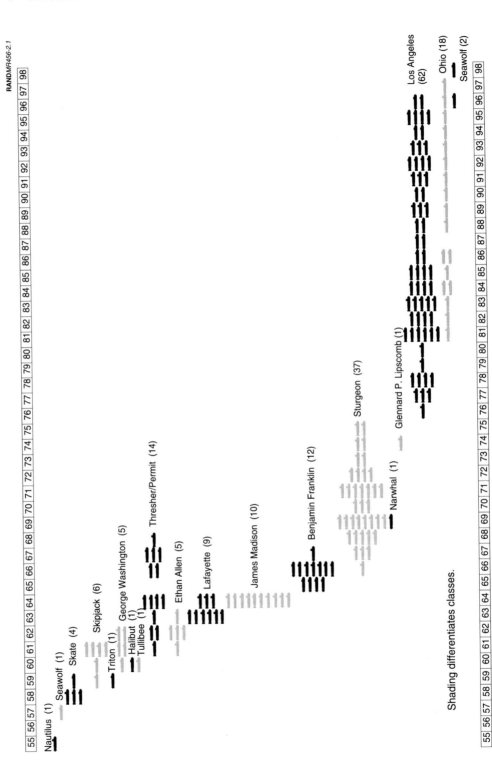

RAND*MR456-2.1*

Figure 2.1—Nuclear Submarines Commissioned, by Class

Meanwhile, the number of diesel submarines was rapidly reduced as World War II vintage subs were scrapped or sold to allies (see Figure 2.2). The Navy's avid pursuit of nuclear-powered submarines through the 1970s permitted the total fleet size to remain relatively constant as the remainder of the diesel-electric ships were removed from service. Now, the submarine force in active service is completely nuclear, comprising about 90 SSNs and 20 SSBNs.

The early nuclear submarine classes were small, often comprising only a few ships; some one-of-a-kind submarines were built. The concepts were new and many unique designs were explored in searching for the best combination of hull form, size, and propulsion and other internal systems. As submarines evolved, class sizes became larger—a result of the early engineering and learning process that discarded unworkable ideas and retained high performance characteristics. Large classes of multimission submarines gained a cost advantage from repeated construction of the same design.

Several classes of submarines are now in service or are being built (see Table 2.1). The current version of SSBN in production is the Ohio class (sometimes referred to as the SSBN 726 class or as the Trident class, after the name of its missile system). Fourteen Ohio-class submarines have been completed, with four more being built. Some older SSBNs of the Lafayette class and the

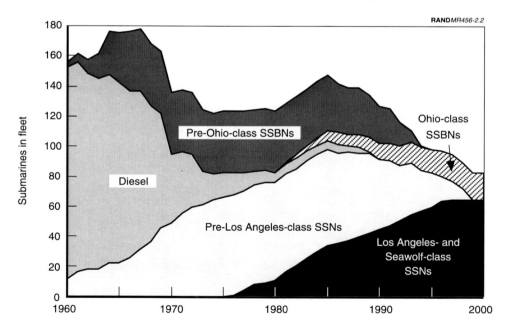

Figure 2.2—Submarine Fleet Composition Profile

Table 2.1

Characteristics of Selected U.S. Submarine Classes

Ship Class	Length/ Diameter (feet)	Displacement (submerged) (tons)	Number Completed (year)	Armament
Gato (SS-212)	312/21	2,391	73 (1941–44)	24 torpedoes
Nautilus (SSN 571)	324/27	4,040	1 (1954)	18 torpedoes
Sturgeon (SSN 637)[a]	292/32	4,640	37 (1966–75)	24 torpedoes and missiles[b]
Los Angeles (SSN 688)	360/33	6,927	39 (1976–89)	26 torpedoes and missiles (for last 8, see 688I)
Improved Los Angeles (SSN 751 or 688I)	360/33	6,927	23 (1988–96)	26 torpedoes and missiles and 12 Tomahawks
Seawolf (SSN 21)	350/40	9,150	2 (+?) (1996–)	50 torpedoes and missiles or 100 mines
Ohio (SSBN 726)	560/42	18,700	18 (1981–97)	24 Trident C4 or D5 missiles and 26 torpedoes

[a]Six of the Sturgeon-class submarines finished between 1971 and 1975 were larger (302 ft, 4960 tons).
[b]A combination of torpedoes and antiship missiles totaling 24.

Benjamin Franklin class remain in the inventory and are scheduled for deactivation in the next few years. Two older SSBNs have been converted to carry out special-forces delivery missions.

The oldest attack submarines in the active fleet are those of the Sturgeon class (SSN 637), whose construction was begun in the early 1960s. This submarine incorporated advanced quieting and sensor systems to make it one of the most successful classes of attack submarines in the fleet's history. In all, 37 of these submarines and one variant (USS *Glennard P. Lipscomb*) were built. Some Sturgeon-class submarines have been deactivated and all remaining ships of this class are scheduled for removal from the fleet by 2000.

In the early 1970s, construction began on the Los Angeles–class (SSN 688) submarine. This submarine was designed with an advanced propulsion plant to give the ship increased speed and maneuverability. In 1980, a major modification was begun with the USS *Providence* (SSN 719) and all following ships (31 in all), in which 12 tubes for Tomahawk land-attack cruise missiles were mounted in the ballast tanks in the bow. In 1983, work was begun on USS *San Juan* (SSN 751), the first of what became known as the improved Los Angeles class. In these 688Is, the forward diving planes were moved from the sail to the bow and the sail was strengthened, allowing ice penetration. Also, the combat system was improved. In all, 62 Los Angeles-class submarines have been authorized, with seven remaining in various stages of construction.

The newest class of attack submarine is the Seawolf (SSN 21), designed to combat the most advanced Soviet submarine threat. It incorporates advances in quieting, firepower, diving depth, sonar, and propulsion. Two Seawolf-class submarines have been authorized and are under construction. The first is scheduled for delivery in 1996.

As for the future, it is anticipated that the SSBN fleet will consist of 18 Ohio-class submarines; all earlier class SSBNs will be decommissioned or modified for other service. Some of the early Los Angeles-class submarines are to be decommissioned early as a cost-cutting effort by the Navy to reduce the size of the fleet. It is unlikely that large numbers of Seawolf-class submarines will be built because of their cost (in the neighborhood of $2 billion). A third ship is planned for a 1996 construction start, though it is conceivable that no more than the two currently authorized will be funded.[3] The Navy is also planning a new SSN class, referred to as the "new attack submarine" (NSSN), more affordable than the Seawolf, to begin construction around 1998.[4] *Whether to build the third Seawolf and when to begin the NSSN are the key issues in defining a production gap for attack submarines.*

Regardless of what start date is chosen, there is no doubt that a new class of submarines will be needed at some point. Observers disagree on the number of attack submarines sufficient to achieve U.S. national security goals in the early part of the next century. But the numbers mentioned within the defense community generally fall between about 40 and 60.[5] As Figure 2.3 illustrates, without any new starts, the number of attack submarines in the fleet will fall below 60 in 2008 and below 40 in 2013, as early Los Angeles-class submarines are decommissioned at age 30.[6]

---

[3]In September 1993, the Administration announced its intention to complete the third Seawolf (SSN-23), and in October, the Department of Defense (DoD) released $540 million in funding previously authorized by Congress for that ship (or other action to preserve the submarine industrial base). Eventual construction of SSN-23 is subject to further congressional funding and approvals.

[4]This program, the planning for which began in 1991, has also been endorsed by the Clinton Administration.

[5]The Administration has endorsed a range of 45 to 55. For a tabular comparison of U.S. and foreign submarine fleets, see Appendix B.

[6]Submarines are designed for a service life of 30 years, although few actually serve that long. The average age at decommissioning is 26. If the Los Angeles–class ships are withdrawn from the fleet before age 30, maintaining the force structure will be a more difficult problem. Service lives longer than 30 years are considered later in this report.

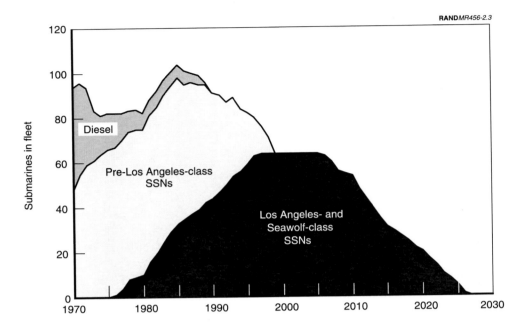

Figure 2.3—Attack Submarine Fleet Composition, Past and Projected,
No New Starts

EVENTS IN THE LIFE OF A SUBMARINE

Time required to design and build, cost to maintain, and when to deactivate are among the factors that must be taken into account in deciding whether to restart production or continue at a low rate. For example, restarting production without an experienced workforce would lengthen design and construction times. Fewer or more maintenance actions may be needed if submarines are decommissioned earlier or later than usual in an attempt to sustain a given fleet size as efficiently as possible. An understanding of the events that take place during a submarine's life is also necessary for an appreciation of the complexity of the tasks involved and thus of the risks entailed in shutting down production and dispersing the workforce. Here, we discuss construction, maintenance, and deactivation of typical attack submarines by an active industrial base. We begin with design, which is not part of the life history of an individual submarine but which must precede construction of the first ship of a new class.

Design

The time required to design a nuclear submarine class has varied greatly. To some degree, design time has been related to technical complexity of the submarine and its systems and the perceived military necessity for fielding the new class. *Nautilus* was designed and engineered in a relatively short time—only three to four years were required to begin construction from the time the Navy made the decision to proceed with the project. The first SSBN experienced an even faster design period. Military necessity demanded rapid development of the SSBN force and the first ship started life as an SSN in the construction phase. During the construction process, systems were altered and the missile compartment was added in order to develop the class more rapidly.

Design periods today (see Figure 2.4) are longer for a number of reasons. As weapon systems have grown in complexity and expense, DoD has instituted rigorous schedules for their review and approval. The acquisition system requires specific milestones to approve the start of concept development, to approve the concept and begin design and R&D, and to approve the design and begin construction. The other major factor contributing to more lengthy design periods is the technical complexity of the submarine. The Seawolf design team had been working for six years before construction started. Extensive investigation was needed to achieve the desired advances in quieting, sonar and combat system capability, and hull form and maneuverability. (For more information on design, see Appendix A.)

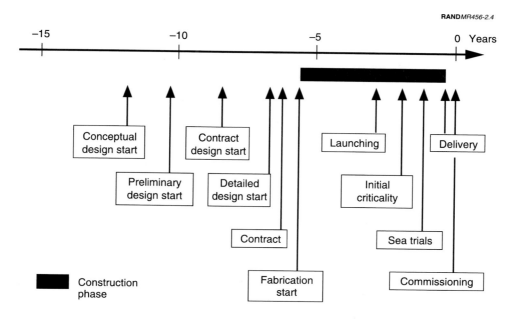

Figure 2.4—Typical Submarine Design and Construction Timeline

Construction

The construction process begins after a contract is awarded to a shipbuilder. The Navy's practice has been to procure long-lead-time equipment (for example, the reactor vessel and other large nuclear components) two years prior to the expected contract award. After contract award, the shipbuilder procures the necessary material and subcontracted equipment and begins fabrication.

In constructing the submarine classes of the past, the hull was erected and openings were cut to enable installation of major equipment. Cramped working spaces and constrained access made running electrical cable, installing machinery, and welding difficult. In addition, one trade had to finish its work before another could start, contributing to construction inefficiencies.

By the early to mid 1980s, the industry had switched to a modular construction approach, in which steel is rolled and welded into hull cylinders and frames. Decks and supports are built into the cylinders, and equipment is loaded into both ends. Completed cylinders of up to 1000 tons are positioned and welded together and the internal piping and wiring are joined. Access difficulties and "waiting time" are minimized, effecting significant cost and time savings.

The submarine is completely assembled inside a building to provide protection from the weather and consistency of welds and measurements. When fabrication is complete, the ship is launched. Once waterborne, all ship's systems are readied for testing. The testing process for a new submarine is complex and technically demanding. Individual components are inspected and tested. Next, systems are tested. In the culmination of the nuclear test program, the reactor plant is filled with coolant, all supporting systems are tested, and the reactor is then operated for the first time and the entire propulsion system is checked out. Likewise, all other systems, including the combat system, are tested while dockside.

When dockside testing is complete, the ship begins a sequence of trials at sea with its eventual Navy crew. These trials test hull integrity, propulsion capability, all sensors and weapons systems, navigation and communication systems, and acoustic performance. The sea trials generally take several months to complete. Once the trials are complete and all material deficiencies are corrected by the shipbuilder, the submarine is delivered to the Navy. This is followed by commissioning, the formal ceremony that officially places the submarine in the Navy's service.

Maintenance

While in service, the submarine operates at sea conducting training, fleet operations, and deployments. During these operations, nearly all preventive maintenance and much corrective material maintenance are performed by the ship's crew or its supporting intermediate maintenance activity (IMA) (submarine tender or base).

A submarine also undergoes a number of shipyard maintenance periods (see Figure 2.5). The first time after delivery that the submarine is detailed to the shipyard is referred to as the postshakedown availability (PSA). Time out of service ranges from 4 to 12 months and serves mainly as a construction guarantee period for the shipbuilder, who repairs and adjusts submarine components and systems that have failed to meet specifications or otherwise require repair. Also, the government normally contracts with the shipbuilder to perform system alterations or modifications that were not performed during construction because of contract cost, time limitations, or material availability.

Another type of maintenance action is selected restricted availability (SRA). SRAs, which are not shown on Figure 2.5, are normally two to three months in duration. The primary purpose of the SRA is to inspect specific systems, which often requires drydocking, and to modify systems and components. SRAs can be accomplished in a Navy or private shipyard, as well as at an IMA.

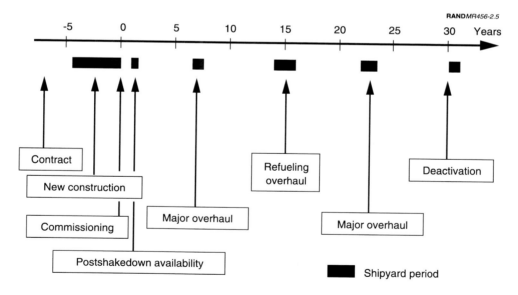

Figure 2.5—Illustrative Submarine Construction and Maintenance Timeline

For the more recently built submarines, the first major shipyard period follow-ing PSA is the depot modernization period (DMP). DMPs normally span 12–16 months. Ship systems are upgraded and major components and systems are refurbished. Inspections requiring major disassembly are also performed (for example, removal of the propeller and shaft to allow inspection of the stern tube).

Overhauls are now conducted once or twice in the life of a submarine. These major shipyard periods can take 18–24 months to complete and take place in ei-ther public or private shipyards. Similar types of modernization, refurbish-ment, and inspections as in a DMP are conducted, but on a more thorough scale. Following an overhaul, virtually all equipment on the ship has been re-furbished. Overhaul may also incorporate refueling. As experience is gained, overhaul frequency and core life are continually reevaluated. Nuclear sub-marines now in service must refuel once. For Seawolf, the core will have suffi-cient fuel and the requisite operating characteristics to last the life of the ship, and that is also the goal for the NSSN. This type of long-lived core will save money by eliminating the need to purchase additional reactor cores and to pay for refueling.

Deactivation

Deactivation removes a submarine from service. The submarine is decommis-sioned—it goes through a formal ceremony striking it from the Navy's list of ac-tive ships—and is taken to a public or private shipyard for deactivation. During deactivation, the nuclear reactor is defueled. All ship's systems are shut down and drained. Fire-control computers and other recoverable equipment are re-moved and sent to other Navy facilities. The ship is prepared for towing and taken to Puget Sound Naval Shipyard. Here, the reactor compartment, which contains all of the submarine's radioactive systems, is cut out of the ship, sealed, and buried at a government site in Hanford, Washington. The remain-ing submarine hull is recycled as scrap metal.

THE SUBMARINE INDUSTRIAL BASE

The prolific submarine-building of the 1950s sustained a competitive industrial base. Seven shipyards succeeded in winning Navy contracts for nuclear-pow-ered ships. Though the capital investment in yard capabilities and expendi-tures for training and maintaining a qualified workforce were great, so were the expected payoffs. As shown in Table 2.2, 209 nuclear-powered ships were built or are under construction. Component suppliers for the nuclear plants and other key submarine systems were similarly busy. The prospect of a large sub-

Table 2.2

Shipyards That Have Produced Nuclear Ships

Shipyard	First Keel Laid	Last Ship	Numbers Built
Electric Boat (Groton)	1952	–	56 Attack submarines[a] 17 Polaris submarines 18 Trident submarines[a] 1 research vehicle
Newport News Shipbuilding and Drydock Co.	1958	–	39 Attack submarines[b] 14 Polaris submarines 11 aircraft carriers[b] 6 cruisers
Electric Boat (Quincy)	1964	1969	2 Attack submarines[c] 2 cruisers
Mare Island Naval Shipyard	1956	1972	9 Attack submarines 7 Polaris submarines 1 Regulus submarine
Ingalls Shipbuilding Corp.	1958	1974	12 Attack submarines
Portsmouth Naval Shipyard	1955	1971	7 Attack submarines 3 Polaris submarines
New York Shipbuilding Corp.	1960	1966	3 Attack submarines 1 cruiser

[a]Including four Ohio-, two Seawolf-, and two Los Angeles–class SSNs under construction.
[b]Including five Los Angeles-class SSNs and two CVNs under construction.
[c]Two additional subs counted here in Groton's total were launched there and towed to Quincy for outfitting.

marine fleet and the promise of nuclear energy led many companies into the production of nuclear equipment.

The boom years of the nuclear shipbuilding industry lasted through the 1960s. Many factors influenced the shrinking of the industry. With lessons learned from the Thresher tragedy,[7] shipbuilding standards became much more rigorous, requiring increased capital investment and sophisticated levels of management and technical supervision. Components had come to last longer, so it was no longer necessary to manufacture as many replacements and spares. Also, the demand for new submarines, which had been averaging eight per year since the late 1950s, suddenly fell to about four annually after 1966, as the

---

[7]USS *Thresher* was lost off New England on April 10, 1963, with its 129-man crew. The only other U.S. nuclear submarine lost at sea was the Skipjack-class USS *Scorpion*, with a crew of 99, off the Azores on May 21, 1968.

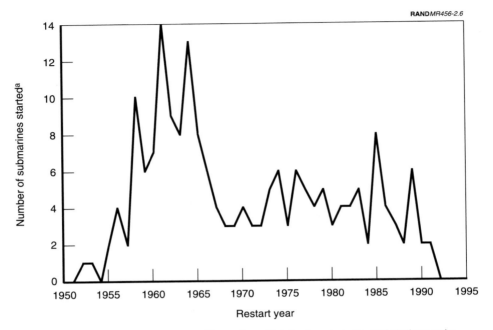

RAND*MR456-2.6*

aNumber of keels laid through 1973; number of fabrication starts thereafter; three subs missed in this transition allotted arbitrarily to 1973 (1) and 1974 (2).

Figure 2.6—Nuclear Submarine Starts Fell Sharply in the Late 1960s, Again in the Early 1990s

Polaris fleet of SSBNs came to completion (see Figure 2.6). Only two yards still construct nuclear ships.

The climate in the civilian nuclear industry worked to the detriment of the component suppliers. Declining orders from civilian plants resulted in many suppliers surviving with only the Navy as a customer. As a result, suppliers of critical nuclear components dwindled from 14 in the 1960s to 4 today.

Currently, there are two private shipyards and six naval shipyards that work on nuclear-powered submarines (see Table 2.3). New construction is undertaken by the two private shipyards—Electric Boat, a division of General Dynamics, and Newport News Shipbuilding and Drydock Company, a subsidiary of Tenneco. Both yards design and construct nuclear submarines, and both conduct activities associated with PSAs and SRAs. Newport News also builds nuclear aircraft carriers, and it can build and maintain other types of ships, both military and commercial.

Table 2.3

Shipyard Capabilities for Nuclear-Submarine Construction and Maintenance

Shipyard	New Construction	SRA	DMP	Over-haul	Re-fuel	Deacti-vate
Electric Boat Groton, CT	X	X	x	x		
Newport News Shipbuilding Newport News, VA	X	X	x	x	x	x
Pearl Harbor Naval Shipyard Pearl Harbor, HI		X	X	X	x	x
Mare Island Naval Shipyard Vallejo, CA		X	X	X	X	X
Puget Sound Naval Shipyard Bremerton, WA		X	X	X	x	X
Portsmouth Naval Shipyard Portsmouth, NH		X	X	X	x	x
Norfolk Naval Shipyard Portsmouth, VA		X	X	X	X	X
Charleston Naval Shipyard Charleston, SC		X	X	X	X	X

NOTE: The 1993 Base Realignment and Closure commission recommended that Mare Island and Charleston be shut down in FY95.

Key: **X**   Currently conducting this operation.
         x   Has capability to perform.

The relationship between the private shipyards and the Navy is handled through contracts. The yards bid on work as solicited by the Navy and a selection is made according to price and other criteria. The contracts are normally awarded through the Naval Sea Systems Command (NAVSEA) and are administered by the Supervisor of Shipbuilding, Conversion, and Repair located at each private shipyard.

The naval shipyards are owned and operated by the Navy. Senior naval officers and government-employed civilians hold supervisory positions in the shipyard organization. The workforce and shop supervisors are civilian employees. Naval shipyards operate under commercial-like procedures, but do not have contracting agreements in the traditional sense with the government. Naval shipyards are administered by NAVSEA.

Naval shipyards no longer construct new nuclear submarines. Although submarines were constructed at Mare Island Naval Shipyard and Portsmouth Naval Shipyard in the past, the building ways are in disrepair and would require substantial capital investment to recover. However, naval shipyards still perform a

wide range of activities from SRAs and DMPs to refueling overhauls and deacti-
vations. Naval shipyards also make emergent repairs during the submarine's
operating life that are beyond the capability of a submarine tender or base. In
addition to on-site activities, shipyards have formed highly trained mobile
"tiger teams" that can make repairs and modifications at Navy bases or other
remote sites.

The private shipyards, component vendors, and naval shipyards represent the
surviving remnants of the industrial base assembled during the late 1950s to
build advanced nuclear-powered submarines that could counter a major threat
to U.S. national security. The early leaders of this effort, and Admiral Hyman G.
Rickover in particular, had an unswerving demand for reliability and safety,
combined with an insistence on personal responsibility. The result was an industrial culture whose record of technical excellence may be unmatched.

Now, the future of the submarine industrial base as a continuing enterprise is in
doubt. Only four new submarines have started construction this decade, and
the last of those is scheduled to be completed in 1998 (see Figure 2.7). If the
third Seawolf-class submarine and the NSSN are not started on schedule, it is
quite likely that the skills built over the years will disperse. In the following
chapters, we analyze the implications of such a loss for the cost and schedule of
submarine production when restarted, and we compare those consequences
with the savings realized from postponing further submarine production.

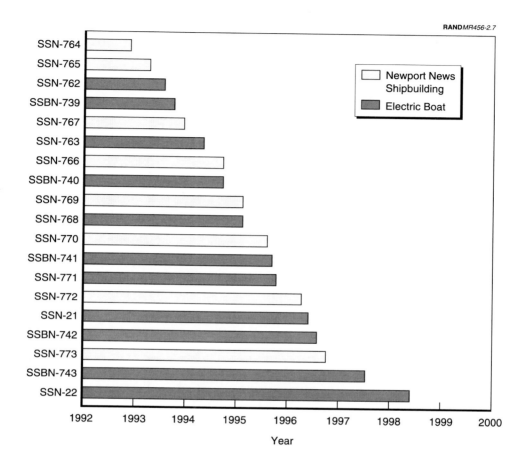

Figure 2.7—Submarine Completion Schedule

Chapter Three

EFFECTS OF A PRODUCTION GAP ON SHIPYARDS

Deferring future submarine production would result in savings arising from a decrease in the present discounted value of that production. It would also result in costs, because personnel would have to be released, facilities would have to be shut down "smartly"[1] and maintained during the gap, and then the workforce would have to be reconstituted. The savings are straightforwardly calculated from the shift in submarine production scheduling, which we take up in Chapters Six and Seven. The costs are not straightforward, and we address them in detail here for the shipyards and in the next two chapters for vendors providing nuclear and nonnuclear submarine components to the shipyards.

We present cost estimates for "smartly" shutting down submarine production capability at the shipyards, maintaining the production capability in a dormant state for a period of time, and then reestablishing production at the shipyards in the future. Postrestart production schedule estimates are given in terms of the time needed to build the first submarine and the time to reach the desired sustained rate of submarine production.

Because the costs of shutdown and reconstitution are greatly influenced by the status of the yards during the halt in submarine production, we examine six distinct cases. In two base cases—one for Electric Boat (EB) and one for Newport News Shipbuilding (NNS)—we assume no new submarine production work at either yard and that the next aircraft carrier (CVN-76) will not be constructed in the foreseeable future. This would leave only the current submarine workload at EB and would result in the complete shutdown of EB's facilities when that workload is finished in 1998. NNS's submarine production work is scheduled to be completed in 1996. Other NNS work includes the construction

---

[1]"Smart" shutdown entails extra investment to preserve institutional knowledge and infrastructure in the interest of promoting efficient restart.

of CVN-74 (*Stennis*) and CVN-75 (*United States*) and the overhaul of CVN-65 (*Enterprise*).[2]

In our third case, we assume the decision is made to build CVN-76 at Newport News with authorization in FY95. This would keep facilities open and allow for a larger workforce at the start of NSSN production than would otherwise be the case. In our fourth and fifth cases, we consider the impact on shutdown and reconstitution costs and schedules for both the baseline cases if submarine repair and overhaul work (including SSBN work) is redirected from the Navy's shipyards to the private yards. Finally, we evaluate a partial EB analogue to the CVN-76 case: constructing a third Seawolf-class submarine (SSN-23).[3]

In this chapter we first discuss certain elements of our approach. We then go through the various cost elements in some detail for the initial baseline case. Summary findings are then given for all the cases considered, along with the principal lessons we drew from them. The other cases are treated in depth equal to that of the initial baseline case (though partially by reference to it) in Appendix C.

METHODOLOGY

This section outlines our general assumptions, our sources of data, and our principal method for modeling the most important category of costs and schedule delays. First, we identify in more detail the costs associated with production gaps at the shipyard: (1) the costs to "smartly" *shut down* and mothball the facilities at the shipyard and to release some portion of the shipyard personnel, (2) the costs to *maintain* the production facilities in a dormant status including the annual costs associated with any personnel retained during the production hiatus, and (3) the cost to *reconstitute* the production facilities and to hire and train the production workforce. Note that each of the three cost components comprises facility- and personnel-related costs. (We also estimate any gap-related costs borne by submarines currently under construction.)

---

[2]We limit our analysis for NNS to submarine and carrier workloads. There is other work currently in the yard, including some commercial ship repair work. And, there is the possibility of other future military and commercial construction or repair work.

[3]EB also builds ballistic-missile-carrying submarines (as has NNS in the past), but replacement of the SSBN fleet will not get under way until after the period covered by this analysis of shipyard reconstitution.

Analysis Assumptions

Assumptions specific to the cases examined are stated in the discussion of results for those cases. General assumptions for the shipyards analyzed are as follows:

- DoD will assume all costs for shutting down, maintaining, and reconstituting the submarine production capabilities.

- Each of the shipyards will be willing to reconstitute production capabilities.

- There will be no political, environmental, or other pressures that would stop reconstitution.

- Nuclear licenses and environmental permits associated with restarting submarine construction will be granted after proper certification by controlling agencies.

- Engineering and design capabilities will be maintained at the shipyard, funded by DoD and Navy research and development moneys.

- Vendors of nuclear and nonnuclear components remain in place.

It is unlikely that all these conditions would be met in the event of an extended gap. If any are not met, costs and delays would be greater than those estimated here. We restrict ourselves in this chapter to quantifiable shipyard production cost and schedule effects. The risks involved in not meeting the conditions listed are discussed elsewhere in this report.

It is important to bear in mind one other matter related to the scope of this analysis. We estimate only costs to DoD. We do not consider any costs associated with loss of employment or reduced revenues to state or local governments during a production gap. Such social-welfare costs are likely to be large.

We also note that only the costs associated with delays in production are included. We do not include costs that would also be incurred if there were no further delay, such as depreciation writeoffs in compensation for capital investment or future workman's compensation and medical claims attributable to the current workforce.

Data for Cost Estimates

Various offices throughout the Navy provided insights and data for this analysis. The majority of our shutdown-and-restart cost and schedule estimates are largely based on inputs from functional-area experts at EB and NNS.

In reviewing these data, we were aware, of course, that both EB and NNS have an interest in the outcome of our analysis. Since EB produces submarines only, a lengthy production hiatus is not in EB's interest. NNS has other options to help it outlast a gap and its principal competitor. We interacted with EB and NNS to fully understand the logic and sources that underlie their estimates. In some cases, we modified the shipyards' estimates based on other data or assumptions. Historical data were also used whenever available. We are satisfied that the data used represent a reasonable, unbiased characterization of submarine production in the shipyards.

Data were also received from several public shipyards on the overhaul and refueling of 688-class submarines. Finally, we consulted with the submarine industries in Great Britain and France to understand how they deal with very low production rates and their experience with production gaps. What we learned from the British and French experience is recounted in some detail in Appendixes D and E.

Modeling Workforce Buildup and Postrestart Production

Methods for calculating most of the cost elements are straightforward and are presented with the detailed results in this chapter and in Appendix C. Calculation of the personnel costs associated with restarting production and estimation of postrestart production schedules are more complicated and are summarized here. Further explanation is offered in Appendix F.

The first ships of a new class built following a production gap will cost more than they would have had there been no gap; they will also take longer to build. Four factors contribute to the extra cost and time:

- The production workforce will have to be rebuilt through hiring and training of workers.

- The rate at which the workforce can be built up will be limited by the availability of skilled workers who can act as mentors.

- In the initial years, the workforce will be less efficient than the preshutdown workforce because it will have a larger proportion of less experienced personnel.

- At the outset, the shipyard's fixed overhead will be spread over fewer ships than it would have been on a steady schedule of two or three ships per year.[4]

---

[4]Fixed overhead per ship is not, strictly speaking, a personnel-related cost. However, the number of ships in the yard is limited by the rate at which the workforce can be reconstituted. We thus

To quantify the cost and schedule penalties of restarting production, we built a model that takes as inputs the initial size of the workforce and the number of experienced workers available in subsequent years from any ongoing production lines for earlier-class submarines or aircraft carriers. Other variable inputs include the number of inexperienced workers a fully skilled worker can mentor and worker attrition rates as a function of experience. Taking all this into account, the model increases the workforce in annual steps, calculating the total worker-hours produced.

In estimating the production schedule, two other inputs come into play: relative worker efficiencies as a function of experience and the target annual sustained production rate. The first of these permits the model to calculate the *effective* number of worker-hours produced per year—that is, the equivalent in fully-skilled-worker-hours. A submarine is "delivered" when enough worker-hours have been accumulated to complete a ship and to get far enough along on others so that progress is being made toward the desired sustained rate. The number of effective hours required to complete a ship is a nominal figure based on experience and anticipation of the character of the next class.

To calculate the cost penalty, the model considers three further inputs: hiring and training cost per worker, worker wages as a function of experience, and variable and fixed overhead.[5] These permit the estimation of total cost over the time it takes to reach the sustained production rate. By subtracting the amount it would have cost to build the same number of ships with a fully skilled workforce at the sustained rate, the total cost penalty is obtained. This is the personnel-related cost of reconstitution.

AN ILLUSTRATIVE CASE: BASELINE ESTIMATES FOR ELECTRIC BOAT

Given the assumptions and approach described above, this section outlines the development of the cost and schedule implications for our first case—Electric Boat becomes inactive after current submarine production ends. We go into some detail here to illustrate the way in which we took into account the full range of shutdown-, maintenance-, and restart-related costs in coming up with the summary results. Similar details regarding the other cases are given in Appendix C.

---

accounted for fixed overhead in our workforce reconstitution cost model, but it was not practical to quantify and separate the portion of total extra cost attributable to fixed overhead (or any other factor).

[5]The model also accounts for the cost of materials, but this was not varied with gap length and is not part of the cost penalty.

Some background on facilities may be helpful in understanding what follows. Groton, Connecticut, is the headquarters for EB and houses the engineering and major shipbuilding functions. The manufacture of complex components, the final outfitting and assembly of hull sections, the loading of nuclear fuel and dockside testing of propulsion plants, and the final operational testing of the complete ship all take place at Groton.

EB's facility at Quonset Point, Rhode Island, has extensive facilities for the automated manufacture and outfitting of major hull sections. Completed modules are barged from Quonset to Groton for assembly into the final product.

For this case, we assume that SSN-22, to be delivered in June 1998, is the last submarine constructed (i.e., there is no SSN-23). This would result in the Quonset facility closing in 1995 and the Groton facility closing in late 1998.

In reporting our cost and schedule estimates for EB, we begin with the additional gap-related cost accruing to submarines now under construction and proceed to shutdown, annual maintenance, and restart. First, however, a methodological point of particular importance to this case deserves some explanation.

Sizing the Residual Cadre

From the remaining workload, we estimated the drawdown in the workforce as submarine construction winds down over the next several years. We assume that the workforce would not be allowed to fall below some threshold number of management and production personnel. This core, maintained during the production hiatus, would be the foundation to build upon and train new workers when production resumes. How large should this cadre be?

We obtained from EB an estimate of the direct production worker-hour requirements, by skill, to produce a notional 10 million worker-hour submarine. The skill workloads were defined by quarter over a five-year construction period. For each skill, we identified the quarter in which worker-hours peaked and converted the worker-hours into approximate personnel levels. The total of the peak-quarter personnel requirements across all skills was 2579.

We assume that at least 10 percent of that number would be needed to form the core workforce to retain during the production hiatus. We therefore base our cost estimates on a core workforce of 260 production personnel. As indirect hours are about 15 percent of direct hours, we add 40 indirect/management personnel to the core to bring the total to 300. We consider this core to be a *mimimum*.

A much larger cadre could be justified. In fact, adding persons to the cadre would result in long-run net savings. The reason for this is that the costs of re-building the workforce after a production gap are large—the smaller the work-force at restart, the larger the cost (this is discussed in more detail below and in Appendix F). The cost of paying an extra cadre worker over the duration of a gap is less than the rebuilding cost saved by having that worker available at restart. With each additional person, the marginal savings at restart decrease, so at some point the net marginal gain drops to zero and a cost-minimizing cadre is reached. First-cut analyses along these lines suggest the cost-minimiz-ing cadre may be several times the size of the one we use.

There is a problem with such an approach, however. Simply proliferating the number of workers on the payroll and the money paid them begs the question of whether skills can be maintained during a gap. What would the cadre do? Over the short run, cadre personnel could provide the labor for the postshake-down availabilities for the new Trident and Seawolf submarines. They also could interact with the design and engineering staffs, working on mock-ups and prototype sections of new submarine designs. Finally, there may be a way to work out agreements with countries such as France, Germany, or the United Kingdom that will still be building submarines during the production hiatus in the United States. Regardless, the larger the cadre, the more difficult it would be to find enough for them to do to retain their mentor qualifications at restart. Thus, we opt for the smaller cadre, although in other cases we do estimate the effects of keeping a larger number of people profitably employed by directing overhaul work to the private yards.

Impact on Submarines Currently in Construction

If production at EB does not restart until FY96 or later, there will likely be some impact on the cost of the eight submarines currently being built. Additional costs would result from two sources. First, incentive bonuses to key manage-ment and trade personnel may be required to ensure that they do not leave the program in the lurch prior to completion of the last submarine by accepting other job offers in anticipation of shutdown. We estimate that at most 2000 personnel would fall into this category and that the average bonus would be $10,000.[6]

Second, one might anticipate a decrease in morale among the workforce. Facing termination of their jobs, there may be a tendency toward decreased productivity. There would also be some loss in efficiency and additional costs associated with transferring work from Quonset to Groton. We assume ineffi-

---

[6]All costs in this chapter (and in Appendix C) are in constant FY92 dollars.

ciencies will increase the remaining workload by a percentage based on when the next new start is authorized. If the next submarine is authorized in FY00 or later, we estimate the total inefficiencies would amount to 10 percent of the 12,000 worker-years remaining. We cost each additional worker-year at $50,000.

Our estimates of the costs are given by year of restart in Table 3.1.[7] If a new start is authorized by FY96, both the Groton and Quonset facilities would remain open and no additional costs should accrue. Later starts result in the closure of one or both facilities, requiring some retention bonuses and resulting in some loss in productivity. The full impact on the cost of submarines now in construction is realized for restarts in FY00 or later.[8]

Costs of Smart Shutdown

Completely shutting down production capabilities at EB will incur costs for facilities and equipment (both nuclear- and nonnuclear-related), personnel release, and vendor liabilities. The first category includes the cost of mothballing equipment to keep it in condition to save time and money at restart. Cranes must be preserved, portable equipment must be palletized and stored, an inventory must be taken, etc. Nuclear-related costs are for securing and monitoring equipment and facilities; we assume there will be no other activities related to handling nuclear components and no environmental cleanup required.

To calculate worker release cost, we assume a workforce profile of 13,400 production and indirect-support employees at the end of the 1993 fiscal year, all but 300 of whom would be released. We allowed $5000 per worker for placement, retraining, and adjustment services (of which $3000 would be for Title III National Reserve Funds for Defense Impacted Workers requested from the Department of Labor). Shutdown-related vendor cost escalation[9] was esti-

---

[7]It is important that this and other similarly constructed tables not be read as expenditure profiles—so much incurred in (or by) FY97, so much in (or by) FY98, and so on. The cost cited for FY97 is the total bonus-and-inefficiency cost incurred in all years if submarine production is restarted in FY97. (Exception: tables of annual maintenance costs.)

Although we do not expect NSSN construction to start before FY98 (and this case assumes no SSN-23), we report results for earlier years. We do this because the cost and personnel data for FY98 and beyond are more easily understood if the results for the antecedent years are also displayed.

[8]The costs of ships in construction could also be affected if the current schedule is either stretched or compressed in response to the NSSN start date. We cannot estimate the likelihood of such an occurrence.

[9]Information from EB indicates that General Dynamics' Quincy shipyard experienced such cost escalation not only during its shutdown but during preceding program downturns that suggested shutdown might be imminent. The extra costs are thought to be the result of several factors, including vendor attempts to recover unamortized development cost over a smaller number of ships and future restart of cold production lines for replacement parts.

Table 3.1

Cost Impact on Current EB Submarines Under Construction

Next Start	Number of Bonuses	Extra Worker-Years	Total Cost (millions of $FY92)
FY95	0	0	0
FY96	0	0	0
FY97	500	300	20
FY98	1000	600	40
FY99	1500	900	60
FY00 or later	2000	1200	80

NOTE: Each bonus costs $10,000; each extra worker-year costs $50,000.

mated at 5 percent of the cost of specialty material and engineered components (the "Plan and Mark" cost) for the last two submarines in production.

The full shutdown costs would be incurred if the next production start is in FY00 or beyond. The pre-FY00 timing of these costs is taken up in Tables 3.2 (supporting data) and 3.3 (costs themselves). For restart before FY00, Groton could remain open; before FY97, Quonset Point could also remain open. Site-specific facility- and equipment-related shutdown costs are timed accordingly. Personnel-related costs are calculated from worker releases based on the anticipated submarine work profile (Table 3.2). Vendor liability costs are assumed to increase as restart is postponed from FY97 to FY00.

Annual Cost of Maintaining Production Capabilities

The longer the gap lasts and the more facilities shut down, the more it will cost on an annual basis to maintain those facilities. Costs will be incurred, for example, for around-the-clock security plus manpower and material for preven-

Table 3.2

EB Personnel Released by Various Dates

Date	Workers Remaining at Start of Year	Workers Released During Previous Year	Cumulative Workers Released
FY94	13400		
FY95	12800	600	600
FY96	9900	2900	3500
FY97	5200	4700	8200
FY98	2000	3200	11400
FY99	700	1300	12700
FY00	300	400	13100
FY01 or later	300	0	13100

Table 3.3

EB Shutdown Costs as a
Function of Next Start
(millions of FY92 dollars)

Next Start	Total
FY95	3
FY96	18
FY97	58
FY98	78
FY99	88
FY00 or later	134

tive and normal corrective maintenance. The objective is to keep expensive equipment and facilities in good working condition to avoid replacement costs.

Utility costs are necessary to cover power to the key buildings for heating, security lighting and alarms, and fuel for the heating plants, compressors, and so forth. Emergency repairs may be needed for unexpected catastrophes such as storms. Other costs include those for taxes, insurance, leases, and pay for personnel in the cadre.

Maximum annual costs are based on historical unit costs and anticipated requirements at Groton. We estimate the cost to sustain the 300-worker core at $80,000 per year per person. Total annual maintenance costs should approach $50 million. This cost applies to gaps leading to restart in FY00 or thereafter.[10] For starts before FY00, some or all of the costs will not be incurred (see Table 3.4), depending on whether Groton or both yards remain open.

Costs and Schedule of Reconstitution

We estimate that reconstituting EB's production facilities and equipment would cost roughly $40 million if all facilities were shut down. These costs include those of opening the facilities, upgrading systems, and setting up equipment. They also include the costs of procedure requalification—obtaining the necessary licenses and other certifications for nuclear operations.

Reconstituting the production facilities and reacquiring nuclear permits and licenses will take several months to several years. However, as will be discussed

---

[10]If production is not restarted until after FY04, substantial repairs of buildings and utility systems may be necessary and some equipment may have to be replaced because of deterioration and obsolescence. As shown in Chapter Six, even relatively small submarine fleets cannot be sustained if restart is put off that long, so we do not address such costs here.

Table 3.4

EB Annual Maintenance Costs Prior to Next Start
(millions of FY92 dollars)

Year	Quonset	Groton	Total
FY95	0	0	0
FY96	0	0	0
FY97	8	0	8
FY98	8	0	8
FY99	8	0	8
FY00 or later	8	40	48

NOTE: Costs on a given line accrue if restart occurs in that year or afterward; for example, if restart occurs in FY97 or thereafter, the $8M on the FY97 line accrues.

in more detail shortly, the limiting factor in getting production restarted will be the time to hire and train the production workforce.

An additional element of equipment-related reconstitution cost is that of re-constituting computers and business information systems and hiring and training computer-related personnel. Because engineering and design functions will continue during any production hiatus, there will be a core computer system to build upon when production resumes. However, that core must be expanded, new software and equipment added, and personnel trained. The farther beyond FY99 that restart occurs, the greater the magnitude of this expansion above the core system.

Personnel inputs to the workforce buildup model are shown in Table 3.5. The first three columns are similar to those of Table 3.2, but only production workers are included. The workers available for restart are 90 percent of those released during the previous year, plus 20 percent of those released the year be-

Table 3.5

Skilled Workforce Available When Production Resumes

Restart Year	Workers Remaining at Start of Year	Workers Released During Previous Year	Workers Available for Restart	Skilled Transfers This Year and After
FY95	10500	500	550	9450
FY96	8000	2500	2350	7200
FY97	4000	4000	4100	3600
FY98	1500	2500	3050	1350
FY99	500	1000	1400	450
FY00	260	240	676	0
FY01	260	0	308	0
FY02 or later	260	0	260	0

fore that. Skilled transfers are 90 percent of those released from current submarine construction lines in the year of restart and thereafter. Thus, for example, in the event of a restart in FY98, we estimate that 3050 workers will be available for restart: 90 percent of the 2500 who were released during FY97 plus 20 percent of the 4000 who were released during FY96. In addition, we anticipate that of the 1500 workers remaining on the 688 and Ohio lines at the start of FY98, 90 percent (or 1350) will become available for transfer to the NSSN line as their work on the current lines ends in FY98 and FY99.

To the costs produced by our model, which considers only production workers, we add the cost of hiring indirect support personnel. We calculate the support personnel needed as 30 percent of the growth in the production workforce after restart and multiply that number by $2000 to account for hiring, relocation, and training. This cost ranges from $4 to $10 million, depending on restart year and production rate.

As shown in Table 3.6[11] and Figure 3.1, the total cost to reconstitute the workforce dwarfs the previous cost elements described. For a sustained production rate of two submarines per year and the other assumptions noted in the table, personnel-related reconstitution costs top one billion dollars if restart is deferred until FY00. Costs escalate as the gap is lengthened or the production rate increased.

Deferring restart until FY00 also increases the time required to build the first submarine from the nominal six years[12] to seven (see Figure 3.2; assumptions

Table 3.6

Total EB Personnel Reconstitution Costs
(millions of FY92 dollars)

Next Start	Rate = 2	Rate = 3
FY98	331	606
FY99	703	1032
FY00	1236	1623
FY01	1665	1972
FY02 or later	1747	2125

NOTE: The case assumes a fixed overhead of $150 million (pessimistic), early attrition rate of 5 to 10 percent (optimistic), and mentor/trainee ratio of 1:2 (intermediate).

[11]Costs in this and analogous tables in Appendix C are expressed with more precision than the estimates merit. These numbers must be combined with the much smaller shutdown and restart costs, so we retain all digits as if they were significant.

[12]In this and other cases, the model yields times of less than six years, but it does not take into account calendar-time requirements for sequential completion of tasks. We assume a six-year minimum for construction of the first submarine. (See Appendix F.)

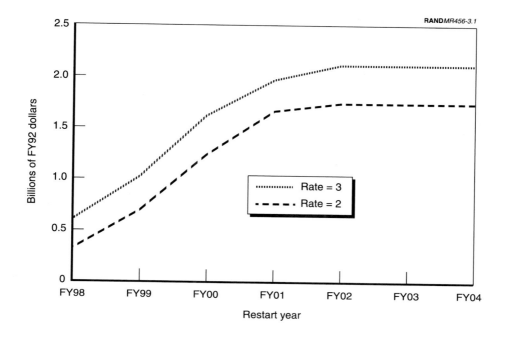

Figure 3.1—Total EB Personnel Reconstitution Costs

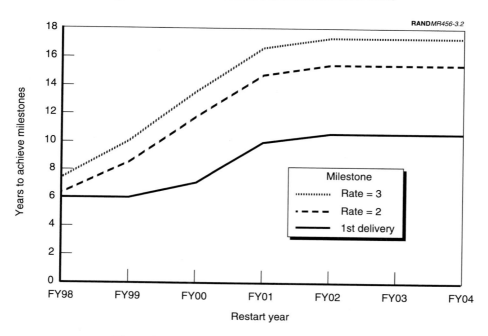

Figure 3.2—EB Postrestart Production Schedule

are those of Table 3.6). Increasing the gap another year increases the construction time to ten years. The time required to reach a sustained production rate of two per year rises rapidly from about six years for an FY98 restart to approximately 15 years for FY01.

SUMMARY OF RESULTS ACROSS ALL CASES

Figures 3.3 through 3.9 and Table 3.7 display the gap-related shipyard costs estimated for each of the six cases identified at the beginning of this chapter (plus one excursion), starting with the illustrative case just detailed. Figures 3.10 and 3.11 show the postrestart schedule results for delivery of the first submarine. Costs are broken out according to the categories identified above.[13] Each vertical bar represents, for restart in the year identified with it, the total extra costs associated with a production gap, summed across all years from the start of the gap into the distant future.

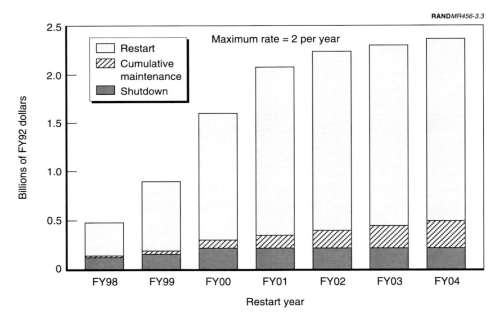

Figure 3.3—Shipyard Reconstitution Costs, Electric Boat, No Work Beyond That Currently Under Way, Maximum Rate = 2 per Year

---

[13]Shutdown costs include impact on submarines currently in construction.

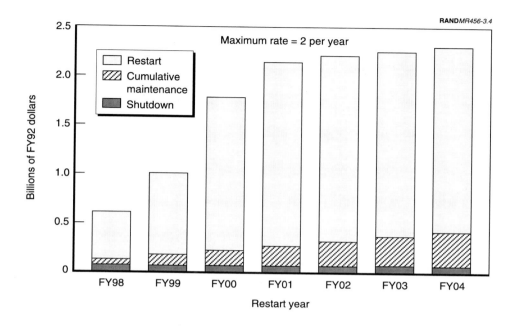

**Figure 3.4—Shipyard Reconstitution Costs, Newport News, No Work Beyond That
Currently Under Way, Maximum Rate = 2 per Year**

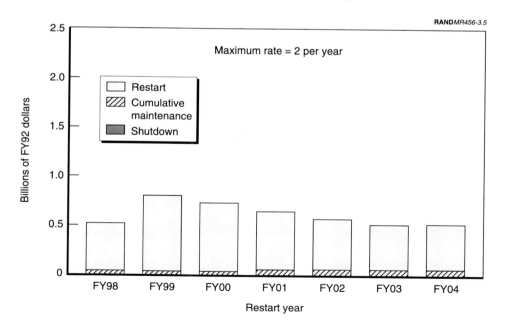

**Figure 3.5—Shipyard Reconstitution Costs, Newport News, with Additional Aircraft
Carrier (CVN-76), Maximum Rate = 2 per Year**

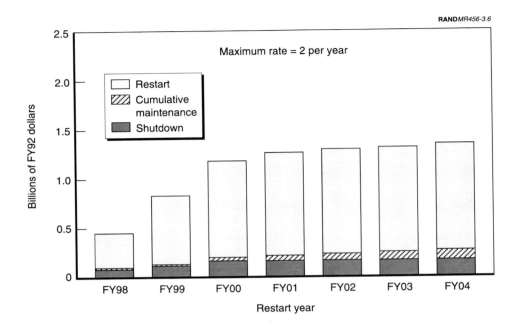

**Figure 3.6—Shipyard Reconstitution Costs, Electric Boat,
with Overhaul Work, Maximum Rate = 2 per Year**

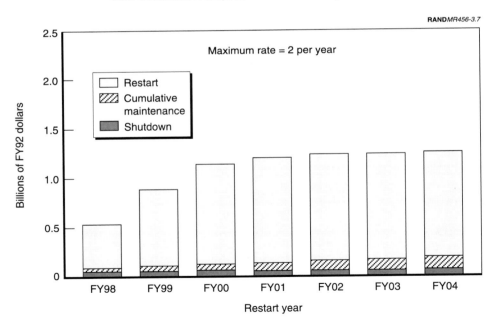

**Figure 3.7—Shipyard Reconstitution Costs, Newport News, with Overhaul Work
(no CVN-76), Maximum Rate = 2 per Year**

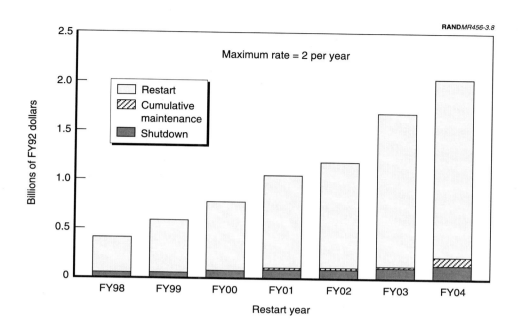

Figure 3.8—Shipyard Reconstitution Costs, Electric Boat,
with Third Seawolf, Maximum Rate = 2 per Year

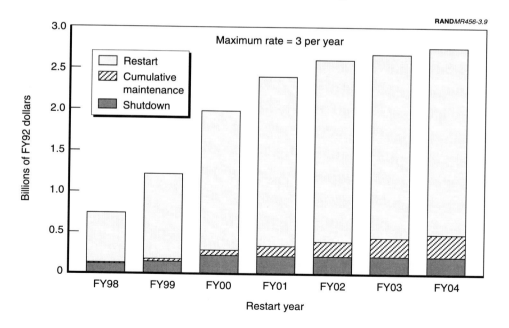

Figure 3.9—Shipyard Reconstitution Costs, Electric Boat, No Work Beyond That
Current, Maximum Rate = 3 per Year

Table 3.7

Summary of Shipyard Cost Effects of Deferring Production

Case	Cost ($M) for Restart in						
	FY98	FY99	FY00	FY01	FY02	FY03	FY04
EB inactive, 2/yr	470	880	1580	2080	2230	2300	2370
NNS inactive, 2/yr	610	1010	1780	2140	2200	2250	2290
NNS w/CVN-76, 2/yr	510	790	720	640	560	510	510
EB w/overhaul, 2/yr	430	820	1170	1260	1280	1310	1340
NNS w/overhaul, 2/yr	540	870	1140	1200	1230	1240	1260
EB w/SSN-23, 2/yr	400	570	770	1040	1180	1670	2020
EB inactive, 3/yr	750	1210	1970	2390	2610	2680	2740

NOTE: Assumptions in Table 3.6 and elsewhere in this chapter apply. For full cost summary, See Table C.21.

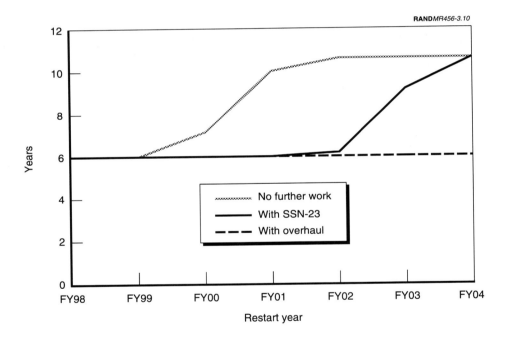

Figure 3.10—Time to Deliver First Ship After Restart, Electric Boat

The results are based on similar assumptions, with some variations (largely workload-based) to allow accurate characterization of the two shipyards. The results shown are based on a sustained production rate of two ships per year. An exception is Figure 3.9, which depicts an excursion from the EB baseline case at three ships per year. Results for the other cases at the higher rate are

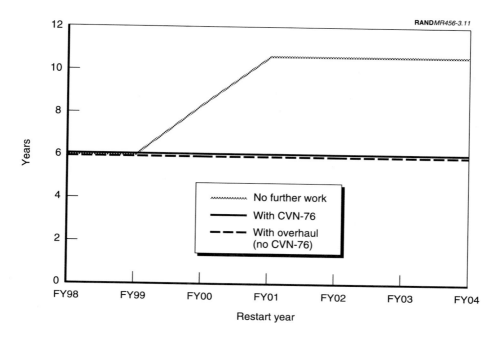

Figure 3.11—Time to Deliver First Ship After Restart, Newport News

given in Appendix C (along with the details of the cost elements supporting the graphs shown here).

The results shown are true only for the set of assumptions they represent (see the note to Table 3.6). We believe the assumptions we made have the effect, taken together, of fairly representing likely cost and schedule effects. The results are thus neither particularly conservative nor anything near a worst case. However, they could change substantially for different reasonable assumptions regarding the rate at which the workforce could be built up, level of fixed overhead, attrition rates, and many other factors (see Appendix F). Nonetheless, absolute values aside, we believe that certain conclusions that rest on the *relations* among elements in the graphs would be fairly resistant to changes in assumptions. The following conclusions represent the central lessons that should be drawn from our analysis of shipyard costs and schedule.

Costs generally increase with gap length. This goes for all three classes of cost elements:

- The longer the gap, the farther the current submarine construction program goes toward completion. Facilities begin to shut down, equipment needs to be stored properly, and personnel are paid severance and retraining allowances as they are released. Eventually, all facilities close and the workforce falls to a minimum cadre that must be maintained to train recruits at restart. Simultaneously, this drawdown could cause inefficiencies in the remaining workforce that increase the cost of submarines currently under construction.

- Shut-down facilities need to be secured and maintained during a gap. The longer the gap, the more facilities need attention and the longer they need it. Also, if a gap is long enough for the workforce to fall to a minimum cadre, then the longer the gap, the longer that cadre must be sustained.

- The more facilities are shut down, the greater the expense to reopen them when production restarts. Also, the more the workforce has decreased, the more it will cost to build it back up again.[14] This is the most important factor in the gap length relations and leads us to our second principal conclusion.

Personnel-related reconstitution costs dominate. This is true across all cases and all restart years. The costs of rebuilding a workforce account for two-thirds to 90 percent of all shipyard reconstitution costs. The reasons for this are given in the factors listed in the description of the workforce model: Not only is it necessary to account for hiring and training, but also for the inefficiency of newly hired workers and the need to allocate fixed shipyard overhead to the few boats that a slowly growing workforce can simultaneously build.

In addition to the delay corresponding to the length of the hiatus itself, **a production gap can add several extra years to the time to construct the first submarine** (which is nominally six years; see Figures 3.10 and 3.11). The main reason for this is that it takes time to rebuild a skilled workforce.

Gap-related costs and delays decrease if other work is available. The equivalent of half an overhaul per year would keep only a small fraction, roughly 10 percent, of the eventual sustained-rate workforce employed. That is enough, however, to cut back on workforce buildup costs and reduce overall gap-related costs—by something on the order of 50 percent from those accruing to an inactive yard (compare Figures 3.6 and 3.7 with Figures 3.3 and 3.4).[15] It is also

---

[14]The workforce buildup model does not account for the difference in skill mix between the residual workforce and that needed to start submarine construction. As explained above, the 260-person cadre can be designed to possess the appropriate skill mix. However, it is doubtful that the skills needed for the early phase of submarine construction would be found in other workers available for restart who are coming off submarine construction lines in the completion phase. Costs for restart in the first few years displayed in Figures 3.3 through 3.9 are thus likely to be higher than the costs shown.

[15]This analysis does not account for the difference between skills and management structures required for overhaul and those required for construction. Although there is substantial overlap (see Appendix C), some skills are construction-specific, so the 50 percent savings should be

enough to eliminate gap-related delivery delays (beyond the length of the gap itself). But it is worth noting that we do not account for the inefficiencies of suddenly directing overhaul work to the private yards and just as suddenly redirecting it elsewhere once submarine production restarts.

A major alternative source of work would have an even greater effect. For example, construction of an aircraft carrier at Newport News (see Figure 3.5) could reduce gap-related costs for some restart years by over 75 percent (and this assumes only one carrier built and no overhauls). As with overhaul, many facilities would remain open and there would be no need to sustain a cadre during the gap. Most important, a great many workers would be available from the carrier line, which winds down over many years, to restart submarine production or transfer into it after restart and thus reduce the dominating workforce buildup costs. In particular, it is the transfers that are mainly responsible for the advantage of the carrier relative to overhaul work, as the initial workforces we assume for both cases are the same for most restart years.

The smaller ongoing source of work represented by the SSN-23 at EB is also (for as long as it lasts) a big improvement over an inactive yard. But it represents a real improvement relative to overhaul only for restarts in FY99 and FY00, when substantial numbers of workers are being released from the SSN-23 line but there are still enough left to provide numerous transfers over the next two or three years.

Increasing the target rate of submarine production increases costs and delays (compare Figures 3.3 and 3.9). A higher production rate does not affect any of the costs involved in shutting down production, maintaining facilities and workers, or reinitiating work. However, a greater number of workers is necessary to build more submarines per year in steady state, and it takes a longer time to rebuild the workforce to that level, so all the costs involved in that effort are increased.

To the costs and delays discussed in this chapter and in Appendix C must be added further gap-related costs and delays—not to mention risks—from having to reconstitute the vendor base supplying the shipyards. It is to these costs, delays, and risks that we now turn.

---

regarded as optimistic. In particular, the initial overhaul would probably take longer and cost more than those performed at the shipyards now carrying them out.

Chapter Four

EFFECTS OF A PRODUCTION GAP ON NUCLEAR-COMPONENT VENDORS

The U.S. Navy has placed enough orders to sustain the naval nuclear propulsion industrial base through FY95. The industry's continued operation depends critically on carrier construction and the next-generation submarine reactor for the NSSN. (The most critical supplier—the supplier of reactor cores—can also rely on carrier and Trident refuelings.) In this chapter, we examine the time, cost, and risk associated with deferring further submarine production long enough that some or all nuclear vendors shut down. The magnitude of the effects raises the question of whether the government should accept the risks and incur the cost involved in letting these firms go out of business and having to restart production at some future date.

This analysis draws on our discussions with and data provided by industry, Naval Nuclear Propulsion Directorate, Naval Sea Systems Command, and OSD officials. Some data come from nuclear-contractor responses to a comprehensive industrial-base questionnaire prepared by the Office of the Assistant Secretary of Defense (Production and Logistics) and administered by the Navy. We begin with an overview of the naval nuclear industrial base.

NAVAL NUCLEAR-PROPULSION INDUSTRIAL BASE

Research, development, and manufacture in support of the nuclear Navy are carried out by major corporations under contract to the government, subcontractors that supply hardware support and technical expertise to the prime contractors, and government laboratories. In this chapter, we focus on the production processes of suppliers who manufacture critical nuclear components:

- reactor cores

- heavy reactor plant components (reactor vessels, steam generators, pressurizers)

- control rod drive mechanisms

- pumps, pipes, and fittings
- instrumentation and control equipment
- valves and auxiliary equipment.

Table 4.1 lists the manufacturers of these components.

The prospects of the naval nuclear industrial base are less than robust. As discussed in Chapter Two, the nuclear industry suffered a reversal of fortunes in the 1970s and 1980s, as is starkly illustrated by the record of civilian reactor orders over the last several decades (Figure 4.1) and the dwindling number of critical-component suppliers to the Navy (Table 4.2). With no domestic civilian orders and declining orders from the Navy, the nuclear field is no longer com-

Table 4.1

Key Nuclear Suppliers

Nuclear Component	Supplier	Location
Nuclear cores	Babcock & Wilcox Co., Naval Nuclear Fuel Division	Lynchburg, VA

Heavy components
  Tubing
  Large forgings
Control rod drive mechanisms
	Babcock & Wilcox
	B&W-Specialty Metals
	Beth Forge
	Marine Mechanical Corporation
		Barberton, OH
		Koppel, PA
		Bethlehem, PA
		Cleveland, OH

Pumps	Westinghouse Electro-Mechanical Division BW/IP International, Byron Jackson Pump Division	Cheswick, PA Long Beach, CA
Pipe and fittings	Tube Turns, Inc.	Louisville, KY
Instrumentation and control equipment	SPD Technologies Eaton Pressure Sensors Division Loral Control Systems GE Reuter-Stokes Peerless Instrument Corporation Imaging & Sensing Technology Corp. Westinghouse I&C Systems Eaton Cutler-Hammer	Philadelphia, PA Bethel, CT Archibald, PA Twinsburg, OH Elmhurst, NY Elmira, NY Baltimore, MD Milwaukee, WI
Valves and auxiliary equipment	Target Rock Corp. Hamill Manufacturing	East Farmingdale, NY Trafford, PA

SOURCE: Admiral Bruce DeMars, *Supplement to the 3 March Report on Preservation of the U.S. Nuclear Submarine Capability*, Naval Nuclear Propulsion Directorate, U.S. Department of Defense, Washington, D.C., November 10, 1992.

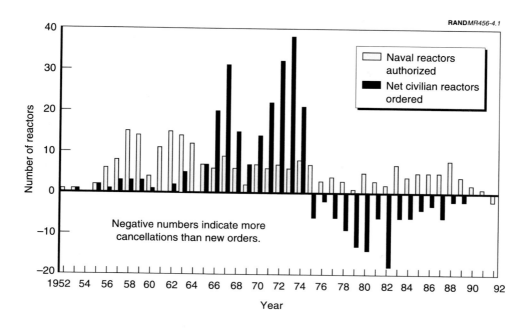

Figure 4.1—Naval Reactors Authorized and Net Civilian Reactors Ordered by Year

mercially appealing.[1] The large capital investment needed and the low probability of achieving a steady return on investment will probably discourage new firms from entering this field. Meanwhile, uncertainty about the timing of future nuclear component orders and anticipated low production rates have raised concerns that the few remaining producers of key naval components may close down or, because of uneconomical order rates, cease making these components.

It is essential for the Navy's nuclear ship programs that the remaining nuclear industrial capability survive. Nuclear system manufacture requires high standards for component manufacturing and quality assurance, specialized facilities for fabrication and testing, and a highly qualified and skilled workforce. Naval nuclear manufacture is even more specialized. Naval nuclear reactors are small, use highly enriched fuel, must operate for decades without replacement or major maintenance, experience frequent power variations, are required to meet quietness and shock criteria, and are designed to operate in close proximity to humans. The means of meeting this kind of demand cannot

---

[1]For a discussion of the commercial industry during its early years see R. L. Perry, A. J. Alexander, W. Allen, P. DeLeon, A. Gandara, W. E. Mooz, and E. S. Rolph, *Development and Commercialization of the Light Water Reactor*, 1946–1976, RAND, R-2180-NSF, June 1977.

Table 4.2

History of Nuclear Component Suppliers

Component	Component Supplier			
	1960s	1970s	1980s	1990s
Reactor cores	B&W UNC CE M&C West.	B&W UNC	B&W UNC	B&W
Heavy equipment	B&W A-C*** AOS FW CE West. Alco	B&W A-C*** C-W CE Aero SW	B&W PCC***a CE SW	B&W
Control rod drive mechanisms	TRW* VARD** M-S	TRW* R/LSI**	TRW* BFM**	MMC*
Main coolant pumps	West. GE	West.	West.	West.

NOTE: *, **, ***-successor company in same facility.

aPCC and B&W consolidated with the eventual outcome of downsizing and PCC exiting from the heavy equipment business.

Key:

A-C	Allis Chalmers		GE	General Electric
Aero	Aerojet		MMC	Marine Mechanical Corp.
AOS	A.O. Smith		M-S	Marvel-Schelber
B&W	Babcock and Wilcox		PCC	Precision Components Corp.
CE	Combustion Engineering		R/LSI	Royal/LSI
C-W	Curtiss-Wright		SW	Struthers Wells
FW	Foster Wheeler		UNC	United Nuclear Corp.
			West.	Westinghouse

be replaced quickly or cheaply, if they can be replaced at all. In the remainder of this chapter, we attempt to convey some sense of the time, cost, and risk involved.

LEAD TIMES WITH AN ACTIVE INDUSTRIAL BASE

Before we discuss the effects of shutting down and then reconstituting the vendor base upon times to produce nuclear components, we offer some perspective by characterizing the lead times now prevailing. As we show, fabrication of nuclear components must begin years in advance of hull construction, even with all production processes up and running—and the lead times have been growing. The advent of modular construction techniques has contributed to this trend. At one time, components were not needed until the hull was finished. Now they are needed to outfit the hull cylinders before these sections are welded together.

Although modular construction techniques are now used on all three current submarine classes, the Seawolf is the first designed for extensive modular construction from the outset. As shown in Figure 4.2, most key nuclear components are required sooner after the shipyard contract award for the first Seawolf than for either the latest Ohio- or Los Angeles–class ships. The NSSN, of course, will also be designed for modular construction.

The intervals between contract award and need date are shorter than the time it takes to manufacture the components. For the SSN-21, fabrication of some components had to start five or six years in advance of the shipyard's work (see Figure 4.3), or twelve years in advance of submarine delivery.[2] Lead times for the NSSN will be comparable. Plans are to order the reactor core in 1996 for a ship that will not be delivered until perhaps 2005. Thus, any delays caused by vendor base reconstitution must be added to these already lengthy lead times.

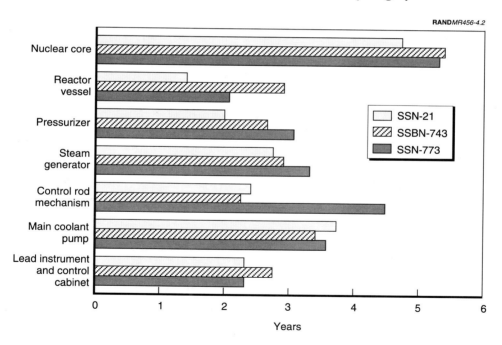

Figure 4.2—Interval from Contract Award to the Time Component Is Needed

---

[2]Because the times shown are for first-of-class ships, they include development and test activities. Times for follow-on ships should be shorter.

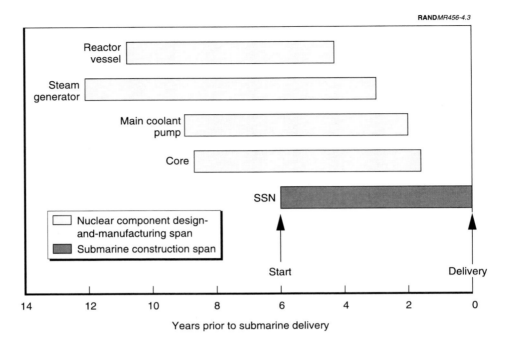

RAND*MR456-4.3*

NOTE: The times assume an active industrial base; required lead times could be longer following an extended production gap.

Figure 4.3—SSN-21 Shipyard Need Dates and Design-and-Manufacturing Spans for Selected Nuclear Components

SHUTTING DOWN AND RECONSTITUTING THE REACTOR CORE VENDOR

We begin our analysis of gap-related costs and delays with the supplier of the reactor core, the costliest and most technically challenging of the nuclear components. We will then examine shutdown and reconstitution effects on the remainder of the naval nuclear industrial base in less detail.

Current plans call for one or two reactor core orders a year from FY95 through FY99. Work on the shipboard version of the NSSN reactor will be preceded in FY95 by start of construction on a prototype. Carrier cores will be started every other year for future refueling requirements, and a Trident refueling reactor will be ordered in FY98.

The core vendor estimates that an annual demand of one reactor core and a half a million worker-hours are required to maintain its financial viability. Planned orders from the Navy are sufficient to meet that goal, but, as production rates are already low and facilities underutilized, the vendor may not remain an economically solvent commercial firm if some of the planned orders do not materialize on schedule. As we show in the following examples, the costs, delays, and risks of losing the core vendor are so great that the only practical alternative to the planned stream of orders is a government subsidy to sustain the vendor's capability.

The following subsections explore two scenarios for shutting down and reconstituting the reactor core production capability. The first scenario examines shutdown of the industry for one year, the second envisions a longer-term shutdown.

The reconstitution process breaks down into five major elements:

- shutting down facilities
- reestablishing the workforce
- reestablishing the facilities
- reestablishing the production process
- constructing and testing prototypes.

For the one-year shutdown, we discuss the first, second, and fourth items; the third should not present a challenge, and prototypes would not be necessary. For the five-year hiatus, all elements come into play.

One-Year Shutdown

We begin our assessment of the one-year shutdown with the effects on timing. Of the three elements relevant to some aspect of the one-year case, only facility shutdown does not affect the timing of reconstitution. In the early years, reconstitution is driven by the need to rebuild the workforce, conduct training, obtain clearances, requalify the production process,[3] and gain preproduction experience. These functions would require about one and one-half years before production begins and continue for about six months during early phases of the production process. The core production process currently requires six to seven years. A one-year shutdown would stretch that manufacturing time by two to three years.[4]

---

[3]The Nuclear Regulatory Commission and the Director of Naval Nuclear Propulsion must certify the safety and quality of new production processes for nuclear components. Processes may have to requalify after shutdowns of six months or more.

[4]This neglects the possibility that the vendor may have to go through a major requalification program (even if the gap in production is as short as six months).

These increased time spans result from reestablishing the production process and reducing uncertainties stemming from the risks of "forgetting" associated with stopping production. While production is in process, manufacturing steps, fabrication methods, and quality assurance procedures are being exercised and yield a known product outcome. In the absence of an ongoing production process, each step will not only take longer to accomplish but also require more frequent checking and component testing (some of it destructive). The checking and testing are necessary to ensure safety and final product quality before proceeding to the next step. This kind of protocol is essential because, in contrast to other high-technology components, a nuclear reactor must work the first time it is used.

Thus, reestablishing the workforce and the production process after a one-year hiatus would together result in an increase of three and one-half to five years in core delivery. This means that steps to reconstitute the core vendor capability would have to begin roughly 10 to 12 years before the core is needed in the shipyard.

A one-year gap would also increase the costs of producing the next core by about $40 million. About half that cost would be for partially decontaminating and decommissioning the facilities, and almost a quarter is for the greater overhead burdens that initial units of production would bear at restart.

Five-Year Shutdown

For relatively long periods of shutdown, costs are difficult or impossible to estimate. Nevertheless, estimates that are available far exceed those for a short shutdown. The total cost to reconstitute would appear to be on the order of a billion dollars, plus an unknown amount for constructing and testing prototypes and first production articles.

For a shutdown period on the order of five years, reconstitution times are even more uncertain. In addition to the usual concerns in restarting any industry—availability of facilities and the ability to hire, train, and retain a work force—the effects of the economic environment, environmental concerns and attitudes, and future licensing requirements are unpredictable. However, assuming all resources are available and today's regulatory environment prevails, as many as 18 years would be required to prepare facilities, develop the subvendors, hire an entire new workforce, train employees, restart production, and produce the first core. Even then, that first core would not be suitable for installation in an operational submarine but would probably undergo rigorous testing as a prototype at a land-based site.[5] This estimate involves a good deal of guesswork, as there is no experience on which to base it. It is also success oriented in that it assumes no major problems develop that require going back to the subcomponent level.

Thus, gaps in nuclear core production for one to five years can double or even triple the time it takes from award of the vendor contract to delivery of the operational reactor. When taken together with the extra costs involved, *it is clear that the core production capability should not be allowed to lapse.*

SHUTTING DOWN AND RECONSTITUTING THE REST OF THE NUCLEAR VENDOR BASE

Prospects for the nuclear core vendor look fairly promising compared to those for the rest of the nuclear vendor base, which does not get as large a percentage of its business from refueling. The work backlog for these firms runs out in 1996, and the NSSN represents the only other foreseeable demand for their services. Deferring reactor component orders beyond the late 1990s will probably mean that critical-component suppliers will exit the business.

We estimate that the capability represented by those firms could be reconstituted more quickly than could that for the reactor core, although reconstitution costs are likely to be substantial. Those for heavy components—reactor vessels, steam generators, and pressurizers—could top $550 million (see Figure 4.4[6]). Fortunately, the backlog situation for these suppliers is the most favorable; design work on heavy components for the NSSN has already started. The costs for the other component suppliers combined come to about $170 million.[7]

RISKS ASSOCIATED WITH GAPS IN NUCLEAR COMPONENT PRODUCTION

Two kinds of risks are not covered in our analysis of cost and schedule effects. One is the risk that our cost and schedule estimates are underestimated—the

---

[5]At present, the Director of Naval Nuclear Propulsion has not proposed prototyping after a production hiatus, should one occur. Such a prototype would either have to replace an existing one now in operation, or a new site would have to be established. That choice would have to be based on difficult tradeoffs involving cost, schedule, technical, and regulatory issues.

[6]The data in this figure are our estimates based principally on an unpublished industry survey taken by the Office of the Assistant Secretary of Defense (Production and Logistics) and adjusted in the light of additional information.

[7]Neither our cost nor our schedule estimates include cost increases and delays in integrating nuclear propulsion plant components and testing their function as a system or in reestablishing coordination among the government, component suppliers, and the shipyards. This coordination, though never perfect, has been the result of many years of working together.

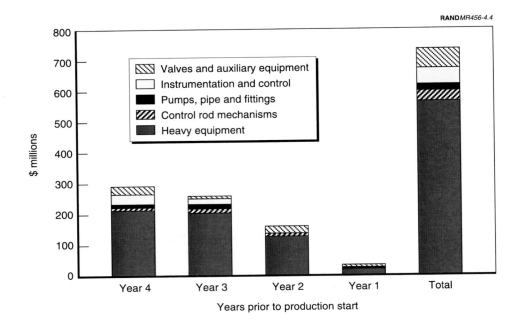

**Figure 4.4—Cost of Reconstituting Nuclear Industrial Base by Year
Prior to Restart**

possibility that the true costs and delays could be so great as to render infeasible the production of quality nuclear components for a near-future submarine
program. The second is the risk that, in attempting to meet cost and schedule
constraints, quality will be compromised and an accident will result.

Estimation Risk

Estimating the cost and schedule of reconstituting the nuclear industrial base is
a difficult and uncertain task, particularly with respect to core production.
Investment programs or projects typically begin with the kind of *initial* or *conceptual* estimates presented here—rough approximations based upon simple
calculations and minimal engineering.[8] Unlike subsequent estimates, initial
estimates reflect the use of simple ratios and factors to estimate basic portions
of the reconstituted facility. These estimates are optimistic and, as mentioned

---

[8]Initial estimates are followed by *preliminary estimates* based upon completed development work
and some engineering and submitted to management for a decision whether to continue the project into plant design. Then come *budget estimates* made when plant design is well under way and
engineering is 30 to 70 percent complete, and *definitive estimates* made when construction is ready
to begin or is under way.

above, success oriented. Usually no attempt is made to estimate the effects of different sites, changing regulatory environments, advances or uncertainty in technology or in the economic environment, and possible changes in scope. In the present case, the relicensing problem alone could add years of public hearings and debates and, at worst, make reconstitution impossible.

Initial estimates have proven to be poor predictors of cost and performance. This is true of various technologically advanced facilities, whether they require advances in the state of the art or not. In past RAND research,[9] data were collected on the accuracy of cost estimates for three nuclear process plants—the Barnwell Nuclear Reprocessing plant, the GE Reprocessing facility, and the Naval Fuels Materials Facility. Barnwell and the GE facility were first-of-a-kind facilities, while the Naval Fuels project was a facility that was to duplicate technology at an aging facility operated by a private firm for the Navy. Figure 4.5 reports initial estimates, preliminary estimates following development work, the actual cost to construct the projects as designed, and the total cost to try to make the projects work to specifications and satisfy regulations until further

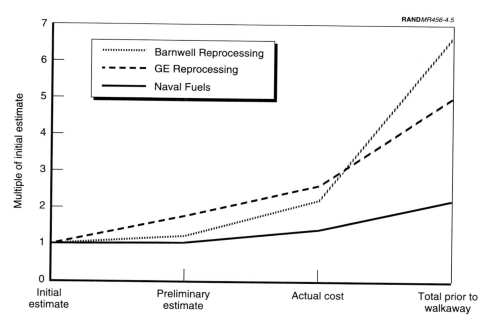

Figure 4.5—Cost Growth for Select Nuclear Facilities (constant dollars)

[9]Edward W. Merrow, Stephen W. Chapel, and J. C. Worthing, *A Review of Cost Estimation in New Technologies: Implications for Energy Process Plants*, R-2481-DOE, RAND, July 1979.

activity was canceled. As is evident from the figure, average increases in cost at succeeding project phases were substantial. The increases shown may also be conservative, for the same reason that it is not possible to report schedule slippage for these facilities: *none of them ever operated successfully.*[10]

For anyone who might imagine that updating a production process technology that has experienced a developmental hiatus is relatively simple, these facilities, especially the Naval Fuels facility, provide some insightful lessons:

- Technologies that are subject to changing regulatory environments must be *reinvented* after any significant amount of time. The Naval Fuels facility was a pioneer manufacturing facility in the truest sense even though it was intended only as a duplicate of an operating facility.

- Design expertise that created and maintained the old technology may not be of much value to the reinvention of the "new" technology.

Thus, it would seem that the initial cost estimates we report here could be multiplied by a factor ranging from two to at least seven (see Figure 4.5), if experience with these sorts of estimates is to be taken into account. However, allowance must be made for the differing incentives behind the various estimates in this chapter. Estimators (including, perhaps, those represented in Figure 4.5) are generally invested to some degree in the continuation of the project they are estimating. They usually do not have strong incentives to account for all possible sources of cost increase and delay in continuing the project. But where the issue is the potential discontinuation of the program, the incentives are reversed. Although we have no reason to doubt the accuracy of the data provided by nuclear vendors, we consider it prudent, particularly in light of the conclusion the data supports, to be conservative in choosing an "underestimation factor." We also note that the duplication of the Naval Fuels Materials Facility is the closest of the three projects in Figure 4.5 to the processes and facilities likely to be undertaken in reconstituting the naval nuclear production base. The growth in estimates for that facility are at the low end of the range.

We thus accept most of the gap costs as stated and apply (in the summary to this chapter, below) a factor of two to the more uncertain longer-shutdown costs associated with the reactor core. We treat the delay estimates, particularly those for the five-year shutdown, as conservative, without attempting to quantify the possible degree of error involved.

---

[10]While the Naval Fuels facility did begin limited production, the high cost of the new facility and declining demand made continued operation impracticable.

Accident Risk

The U.S. Naval Nuclear Propulsion Program strives to reduce risk to minimum practicable levels. This philosophy was imbued in the program by Admiral Rickover from its inception in 1948 until 1982.[11] Rickover dominated the development of the program and was personally responsible for its content and effectiveness. He had an uncompromising attitude toward safety, rejected highly developmental and untried systems and concepts, and instilled a strong safety culture in the Navy's reactor program. Accidents were—and are—unthinkable. His approach to design was conservative and required strict control over manufacture of all equipment, including extensive inspections by specially trained personnel during the course of manufacture and of the finished equipment. Safety during the handling of nuclear material was and is absolutely paramount. As a result, there has not been a single nuclear accident in the history of the nuclear navy.

While we cannot quantify risk, we have drawn on available literature to set up a framework that allows some qualitative inferences about the future. This literature suggests a set of eight parameters experience has shown are significant indicators of decreased accident likelihood:[12]

1. Absence of new or unusual materials

2. Absence of new or unusual methods of construction

3. Absence of new or unusual types of structure

4. Experience and organization of the design and construction team

5. Sufficiency and relevance of research and development background

6. Favorable industrial climate

7. Favorable financial climate

8. Favorable political climate.

With respect to the first three criteria, the current overall approach has been conservative, averse to unproved technologies, systems, and materials. The

---

[11]Upon Rickover's retirement, Executive Order No. 12344 established the position of Director, Naval Nuclear Propulsion, at the four-star Admiral level. It vested in that position direct personal authority for all aspects of the nuclear propulsion program—operations, safety standards and procedures, research, etc. That authority is entrenched by the security of an eight-year term in the post. The order recognized the prerequisites for maintaining a rigorous safety culture by institutionalizing for Rickover's successors his power to dictate the shape of the program free of interference from other branches of the Navy and government.

[12]Alfred Pugsley, "The Prediction of Proneness to Structural Accidents," *The Structural Engineer*, Vol. 51, No. 6, June 1973.

quality of reactor and equipment design and manufacture are confirmed through extensive analyses and full-scale mockups and tests using land-based prototypes. The approach has been deliberately evolutionary.[13]

With regard to criterion 4, the current design and construction teams have considerable experience. Scores of naval nuclear reactors have been safely built and operated over a period of decades. As for 5, land-based prototypes of design similar to that of shipboard plants are built first and operated continuously, so that any design-related problems should be noticed at a land-based site before they occur at sea. During the past decades, the submarine community has enjoyed high levels of business, financial, and political support.

It would appear that criteria 1, 2, 3, and 5 will continue to be met in the foreseeable future. Criterion 4 would, of course, not be met if production is suspended. The design, development, and construction of nuclear submarine components is not a theoretical skill learned in textbooks. Rather, these skills come about through experimenting with methods and ideas, testing them, and learning through trial and error. Newly formed teams lack these experiences and must learn by making a few mistakes of their own. Obviously, if not caught, mistakes heighten the potential for accident.

While the latter is the most important potential shortfall, the submarine industrial base may not see the kind of business, financial, and political support after a production shutdown that it once drew upon to start a high-quality program. Thus, according to our qualitative framework, at least four of the eight elements are likely to foster risk, a situation counter to the philosophy that has underpinned a highly successful endeavor. *Unless a costly, time-consuming effort to reduce risk is undertaken, the potential for accident can be expected to increase.*

SUMMARY AND CONCLUSIONS

Shutting down and reconstituting the naval nuclear industrial base will result in increased costs, longer contract-to-delivery schedules, and an elevated risk of deficiencies in product quality and thus of accident. Attempts to quantify these effects are fraught with the potential for error. With that caveat in mind, we offer the following conclusions:

- Deferring orders for the NSSN or additional submarine reactor cores beyond the late 1990s could result in inefficiencies at the core manufacturer that would translate into higher overhead costs, which would be borne by the government.

---

[13]Francis Duncan, *Rickover and the Nuclear Navy: The Discipline of Technology*, Naval Institute Press, Annapolis, Md., 1990.

- Deferring production of the NSSN beyond the late 1990s will probably result in the shutdown of the capability to produce other critical nuclear components. Reconstituting this capability after an extended gap may not result in long delays but could entail costs in the hundreds of millions of dollars. Some critical-component vendors could exit the business within the next year or so; the government must begin taking action soon if large costs are to be avoided.

- Because the increase in accident risk following a production gap is unquantifiable, we cannot say whether it would be significant. However, it will be greater if components are pushed through the production process on a budget and schedule much tighter than those projected above. While the probability is uncertain, the consequences of a submarine-related nuclear accident would clearly be severe. An accident could well entail loss of life if it occurs at sea, greater loss of life and social and environmental disruption if it occurs in port, and the suspension of the nuclear submarine program in either case.

Chapter Five

EFFECTS OF A PRODUCTION GAP ON NONNUCLEAR-COMPONENT VENDORS

A major element of the overall submarine industrial base is the set of vendors that supply nonnuclear components and material to the shipyards. There are roughly one thousand firms that provide components of some technical complexity or that are unique to submarines. The majority of those products represent variants of products used elsewhere and do not require highly specialized industrial processes. However, about 10 percent of the products are distinctive to some degree, and some are highly specialized and provided by firms that have no other product lines. If new submarine production were to be stopped for a prolonged period of time, many of those specialized product lines would be abandoned and at least some of the firms would go out of business. The resulting loss of capability would have to be replaced before submarine production is restarted, resulting in extra costs and delays. In this chapter, we (1) quantify the number of products or firms that would be most affected by any prolonged cessation of submarine production, and (2) define and provide a preliminary evaluation of two strategies that might be adopted to provide an adequate supply of such critical components when submarine production is restarted.

WHY IS THIS AN IMPORTANT ISSUE?

Why does a submarine require components and material much different from that of any other naval vessel? The answer is found in several characteristics of submarine operations:

- Pressure. Along with the hull itself, some of the components are exposed to the water pressure created by deep submergence. The propeller drive shaft, periscope, and other items for which there are hull openings must remain sealed against sea water. The steam condenser in the propulsion system and other parts that circulate sea water for cooling must operate at deep-submergence pressures.

- Noise-limiting. Pumps, motors, gears, and other mechanical devices must be specially designed to minimize the noise they produce.

- Internal environment. A submarine must create and maintain an underwater life-support environment that provides comfortable and healthy conditions for the crew and for sensitive equipment.

- Reliability. Because of the long operating tours and the catastrophic consequences that can result from equipment failure, high standards of reliability are demanded from many components.

While almost every component of a submarine has some counterpart in surface ships, the combination of the above special subsurface operation conditions requires unique design and performance specifications for some items. Simply because a generic class of components, such as hydraulic pumps, is widely available does not guarantee continuing availability of similar components suitable for use in combat submarines.

In addition to the special requirements that many of the components must meet, three other characteristics of the nonnuclear supplier base contribute to the difficulty of ensuring supply after a production gap. In many cases, the product was developed by the firm using at least some of its own resources, and thus the firm retains proprietary rights to the product. Furthermore, many of those designs have been refined over the course of several generations of applications, some dating back to the earliest nuclear submarines. Experience with those earlier designs, together with increasingly stringent performance specs, has in at least some cases led to a product that appears to be relatively simple but in fact requires highly specialized and subtle manufacturing equipment and processes. Finally, because of the dwindling demand for submarine-specific products, in many cases there is now a single qualified supplier for a component, making it more likely that at least one source will outlast a gap.

SCOPE OF THE PROBLEM

The first objective of our nonnuclear-vendor analysis was to identify and quantify the components and suppliers that potentially would become unavailable in the event of a prolonged period of no new orders for construction. We turned to three data sources:

- A survey and analysis of the submarine industry conducted by the Navy, initially in early 1992 and then updated in November 1992.[1] Among other

---

[1] *Preservation of the Industrial Base for Nuclear-Powered Submarine Systems: Fall Update*, Office of the Assistant Secretary of the Navy for Research, Development, and Acquisition, November 1992.

topics, that report contained results of a survey of 233 vendors conducted by the Naval Sea Systems Command's Shipbuilding Support Office (NAVSHIPSO).

- Lists of suppliers provided by the shipyards, and ranked in terms of criticality (by the shipyards' own criteria).
- Visits to a few of the supplier firms by members of the RAND research team.

RAND drew on data contained in the NAVSHIPSO survey to develop a screening process and to make a first-order determination of critical vendors. NAVSHIPSO identified 114 of the firms returning surveys as posing a potential problem of continuing product availability after a production gap. We restricted our analysis to those firms. Because of the limited scope of the source data, our analysis is not exhaustive, but we believe it to be complete enough to support a useful set of conclusions and recommendations.

Because the survey sample contained information on a wide variety of products and firms, our next step was to screen the members of the list in a way that would sort out the most critical ones. We defined a two-step screening process:

1. Will the firm's product still be available in the absence of new submarine production for a period of several years? If so, the product is not critical; if not available, then ...

2. Will the product have been technologically superseded after a production gap of several years? If so, the product is not critical; if not so, it is.

Application of these criteria screened out most firms. In some cases the firm's product, while special to submarines, used standard production processes and constituted at most a special production set-up. An example is the special steel used in the hull. That steel requires a special mill run, but is produced on equipment used to produce many other grades of steel for a wide variety of customers. Products such as these were screened out in the first step. That is, we would anticipate that at submarine production restart, they could be available with little delay or extra cost.

Alternatively, a product might be evolving so rapidly that a new design is produced every few years, thus outmoding the existing design. We believe that all of the electronic mission equipment such as sonar and communications gear falls into this category. Both the hardware and software elements of such equipment are evolving rapidly so that any future submarine would probably use designs substantially different from those used in today's submarines. Furthermore, the industry that supplies electronic mission equipment is broad-based and relatively robust, with many related products being produced for surface ships, the aerospace industry, and other military applications. While we cannot assert that future submarine components of this class will be "easy" to develop and produce, we see little basis for concern that vital elements of the industry capability will atrophy in the short-term absence of new submarine orders.

After application of both screening tests, 11 products were judged critical. These were components that might still be technologically current but for which the production line would close and the supplier might go out of business over a prolonged gap in new orders unless some mitigating action was taken.

The next step was to compare our list with similar lists provided by each of the two shipyards. Expecting a high degree of correlation, we found instead little agreement among the three lists. Out of the 53 products identified in the various lists, only three appeared on all three lists, and only a total of 12 appeared on at least two of the lists (see Table 5.1).[2]

The lack of agreement on identifying critical products results partly because of differences in the rules used by the three organizations in the analysis. However, this preliminary exercise was sufficient to justify two important conclusions regarding the submarine vendor base:

- The vast majority of vendors are expected to remain as viable suppliers even across several years of no new submarine starts. The number of products and suppliers that require some significant government action to ensure continuing product availability is, at most, a few dozen. This conclusion becomes less certain as the next submarine start is delayed into the late 1990s, and would be expected to change substantially if the next start is delayed until 2000 or later.

- A few production lines will probably be shut down before 1995 because they supply items needed early in submarine construction and the order gap is already reaching critical proportions for those firms. Action must be taken immediately if these industrial capabilities are not to be lost.

POSSIBLE ACTIONS TO ENSURE FUTURE PRODUCT AVAILABILITY

For cases in which production lines may be shut down, the Navy is faced with two broad choices—preserve the present firms and lines or allow those lines to close and reconstitute a source of supply in the future. Each of those options is briefly explored below.

---

[2]In addition to differences among the three "short lists," three of the firms in Table 5.1 did not appear on NAVSHIPSO's "long list" of 114.

Table 5.1

Nonnuclear Suppliers and Components Judged Critical by at Least Two Sources

Firms rated critical by all three evaluators

Cepeda Associates (air scrubbers)
IMO-DeLeval (steam condensers)
EG&G Sealol (main drive shaft seals)

Firms rated critical by two evaluators

Allied Signal/Garrett (hydraulic pumps)
CBI Services (sonar spheres)
Fairbanks Morse (diesel generator sets)
Hitco (sonar domes)
Martin Marietta (towed-array capstans)
SPD Technology (switchgears, circuit breakers)
Vacco Industries (air reduction valves)
Waukesha Bearings (main drive thrust bearings)
York International (air conditioners)

Preserving Existing Production Lines

Several vendors discussed with us how their submarine-specific capabilities might be preserved across a production gap of several years. From these discussions we made a list of general options. Although no attempt was made to work out a specific strategy for a specific firm, it seems clear that enough different strategies exist so that any of the critical production lines could be sustained for a few years if that was deemed appropriate.

By examining these strategies we are not necessarily recommending that any particular product be "saved." Such a decision should be reached only after developing a strategy for that product, estimating the cost of applying the strategy, and comparing that cost with the expected cost and risk of establishing a new supplier at some future time. Analysis at that level for each of the several dozen potentially critical products was beyond the scope of this study.

We identified five general strategies that might be applied if a decision is made to preserve a production line. All, of course, entail extra costs of some kind to DoD, and these costs need to be weighed against the benefits.

First, for those products that require overhaul or replacement during normal use, there is some continuing demand for activities that could be met by the original supplier. In some cases, that supplier is now doing the support work, thereby sustaining the firm's critical staff, paying overhead on the facilities, and so forth. However, in many cases that work has been contracted to other firms or is being done within the Navy's own facilities. Shifting that work back to the original supplier would not be cost free—someone else would lose the work—but it might well prove to be the least expensive solution.

Another option is to purchase, lease, or otherwise retain key facilities required to manufacture the critical product. The manufacture of the submarine-specific products often requires unique tooling and test facilities that cannot be used in a firm's other products. By relieving the financial burden of preserving those facilities through an idle period, it might be possible to ensure quick restart in the future.

A third option is to purchase from the supplier of the critical component a quantity of related products used in surface ships that is sufficient to sustain that supplier. Again, that means a loss of work elsewhere, presumably on the part of a supplier that is charging lower costs or providing higher quality for the surface ship component.

A fourth possibility might be to purchase additional quantities of the latest versions of critical components and use them to upgrade older submarines during overhaul. Given the reduced threats of the post–cold war era, it seems unlikely that this strategy would be chosen.

Finally, the gap could be used as an opportunity for the firm to enhance its production processes, improve quality and reliability, and reduce costs.

Reconstituting a Source of Supply in the Future

If submarine construction is not expected to be restarted for several years, it becomes increasingly attractive to simply allow the present production lines to be closed, in the expectation of "starting over" when the need arises. In some cases that might mean asking the original supplier to re-create the production facility and then requalify the process. In other cases it could mean going to a supplier of related products and persuading that firm to create the needed product design and associated production facilities. We attempted to determine the costs and risks of starting over, to give some sense as to whether this might be a better approach than preserving existing lines. As will be apparent from the discussion that follows, that attempt was beset by a variety of difficulties.

Cost of Reconstituting an Old Production Line. We used two data sources to estimate the cost of reconstituting the current supplier base. The first was a series of estimates, made by the Los Angeles– and Ohio-class submarine vendors, of the cost of restarting their own production line after a production gap of unspecified length. That estimate was provided as part of the industrial base survey conducted by NAVSHIPSO in the first half of 1992. The second data source was the production cost per shipset for several dozen of the most critical components.

The vendor-estimated cost of restarting a production line was available for only some of the critical components. Furthermore, that cost, and the cost of the components themselves, varied widely. To permit some comparison among the components, we determined the ratio of nonrecurring restart cost to recurring unit production cost for a dozen components where both values were available. Most of the ratios fell between the values of 1.0 and 1.5. That is, the suppliers estimated that the nonrecurring cost of restarting an inactive production line would typically be slightly more than the unit production cost of one shipset. Recognizing that these restart cost estimates had been provided by the suppliers themselves, without any critical review, we chose to use the (optimistic?) value of 1.0.

We also determined the total unit production cost of one shipset for 20 of the most frequently mentioned products in the various listings of critical items. That value is about $100 million. Multiplying that by the ratio determined above should give the cost to reconstitute all 20 production lines.

The number needing reconstitution will depend on the length of the gap. Some firms will complete their current production runs in the next year or so, while others have active production of closely related products (e.g., propulsion reduction gears for surface ships) scheduled for several more years. We estimated that if production was restarted in 1995 or 1996, only about one half (by shipset value) of the producers would have to reconstitute their production line. In contrast, if submarine production was delayed until the late 1990s, all of the critical suppliers would incur reconstitution costs.

Thus, we estimated that for a near-term restart of submarine construction, the cost of restarting production for the critical vendors might be on the order of $50 million, while further delay could double that cost. In all cases these estimates refer to restarting existing vendors' production lines, and assumes that in most cases critical production facilities and test equipment have been retained during the production gap.

The estimation process outlined above is obviously rough and subject to considerable uncertainty. We have deliberately shaded the estimates toward the low side. As discussed in Chapter Four, early, rough cost estimates have a tendency to grow. Taking that into account, we believe that the actual costs of reconstituting production of critical vendors could easily be twice the stated values of $50 million to $100 million even if submarine production is restarted before the end of this decade.

Cost of Developing a New Supplier. We discovered wide variation in estimates of how difficult it would be to qualify a new supplier for many of the critical parts[3]—ranging from strong pessimism to strong optimism about the practicality of doing so. The variation stems from a lack of generally available data. As mentioned above, critical products have usually been designed and developed by the producing firms, who then hold proprietary rights in the process and are reluctant to release information. Thus, how much the current producers spent in developing their products is unknown. This is complicated by the evolution of those products to increased performance levels over the intervening decades.

Limited research suggests that the nonrecurring development cost of complex military hardware is typically several tens of times the unit production cost. Even a multiple of ten, lower than anything in the available database, would suggest that qualifying new suppliers for the 20 critical components noted above would cost at least one billion dollars. Using the past as a guide, we would expect that the suppliers would be willing to provide some of that investment. However, it also seems plausible that, given the turbulence of such markets during the 1990s, few suppliers would be willing to invest the full amount needed. Thus we face two uncertainties—the actual cost of qualifying new suppliers, and the fraction of that cost that would be borne by the government.

Let's assume that, on average, the government would have to provide half the cost of creating a new supplier base. In accordance with the second sentence of the preceding paragraph, let's further assume that the total cost would be only ten times the unit production cost. These assumptions yield an estimate of half a billion dollars. We believe that is the minimum that would be required as government investment, while the actual cost could easily be several times that amount.

Conclusion. In addition to the estimating difficulties already discussed, the accuracy of our estimates must be discounted to some degree because we were not able to take into account a variety of factors, such as

- uncertainty in the design specifications of the next submarine

---

[3]We assume that a new supplier would have to start from scratch. It is widely believed that it would be impractical for the government to simply buy the rights to design-related information from an exiting producer and provide it to a new one at restart. Manufacturing and testing practices and procedures vary so much from one firm to another that it is generally easier to redesign a product than to adapt existing production processes.

- the degree that industry is supported over the intervening years by the construction of surface ships, including aircraft carriers[4]

- the future acceptability of overseas suppliers

- the status of the overall economy and the quality of the industry infrastructure when restart occurs.

Because of the various problems in estimating reconstitution costs, any cost projection would be uncertain and subject to the risk of substantial underestimate. As a result, we conclude that DoD should approach such a process with caution. We believe that both the cost and the risks of starting over will almost certainly be larger than the costs and risks of sustaining the present suppliers for a few years.

COMBINING SHUTDOWN, MAINTENANCE, AND RESTART COSTS

We combined the costs estimated in Chapters Three through Five for various production restart strategies. An example is shown in Figure 5.1. These costs were then added to the costs of sustaining the fleet, discussed in the next two chapters.

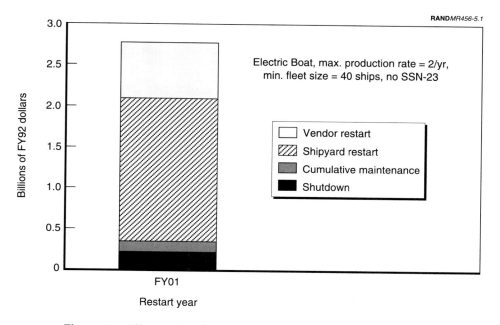

Figure 5.1—Illustrative Shutdown, Maintenance, and Restart Costs

_____

[4]Independent of any attempt to sustain current suppliers by redirecting work.

Chapter Six

ALTERNATIVE FLEET REPLACEMENT STRATEGIES

We have estimated the extra costs associated with a submarine production gap and the delays caused by having to reconstitute the workforce and production facilities for restart. In Chapter Seven, we will combine those cost elements with others whose deferral results in gap-related *savings*, to compare overall costs for production gap strategies. First, however, we need to define a mechanism for examining the cost and schedule consequences of each strategy. That will lead us to defining the strategies—combinations of production rate, fleet size, and ship life—that are practical alternatives for evaluation from the point of view of basic standards of feasibility, national security, economics, and analytic efficiency.

MODELING FLEET REPLACEMENT

Although the elements of the strategies we've just been talking about all affect gap length, their scope and implications are really much broader than that, so we prefer to call them "fleet replacement strategies." To determine the outcomes of different strategies, we constructed a linear-programming model. The model is designed to choose a fleet replacement schedule, including a restart date, that yields a minimum net present value (NPV) of future costs over a designated time period, subject to specified constraints that characterize the strategy.

The overall structure of the model is outlined in Figure 6.1, with constraints listed on the left. The following ranges of values were investigated:

- **Sustained fleet size:** 30, 40, 50, or 60 ships. These choices bracket the current force size goal of 55 established by the Joint Chiefs of Staff. The use of the other numbers is not an endorsement of lower goals but stems from our desire not to bias the results against an extended gap by ignoring fleet sizes that permit such a gap.

RAND*MR456-6.1*

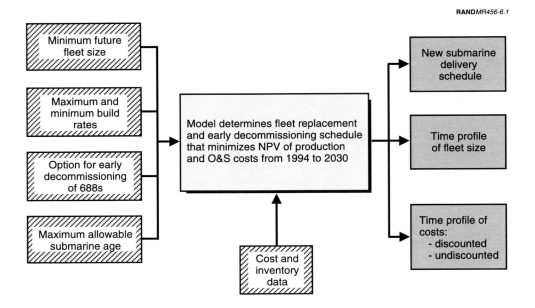

Figure 6.1—Fleet Composition Analysis Framework

- **Annual production rate:** a *maximum* of one, two, or three ships per year, with one or two active shipyards, and a *minimum* of zero, one half, or one ship per year. Can the shipyards deliver at the maximum rates specified? Figure 6.2 shows that Electric Boat can, and Figure 6.3 shows that Newport News can deliver two per year and has at times delivered more.[1]

- **Lifespan of current ships** is investigated in the form of two choices:

  — Whether to allow early decommissioning of Los Angeles-class ships so as to reduce operating and support (O&S) costs in the near term when the fleet size is greater than is justified by the present national military strategy and the threat it assumes. Decommissioning is considered only at the designated refueling point (around age 16 for the early models, and at about year 24 for the later models[2]). Neither Seawolf

---

[1]To depict full submarine production capabilities, these figures include past SSBN construction. Future SSBN production is likely to influence SSN construction scheduling and costs. However, there are too many uncertainties about the nature of the future nuclear threat and the appropriate SSBN fleet size to attempt to account for SSBN effects.

[2]This was the information available to us in the spring of 1993. The Navy has since informed us that reactor cores on some later-model 688s will have sufficient fuel to last beyond year 24; where they do not, the ship will be decommissioned.

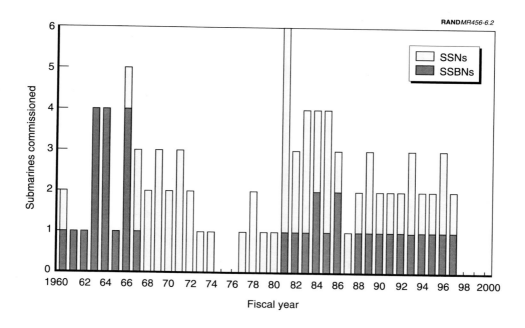

Figure 6.2—Electric Boat Submarine Production History

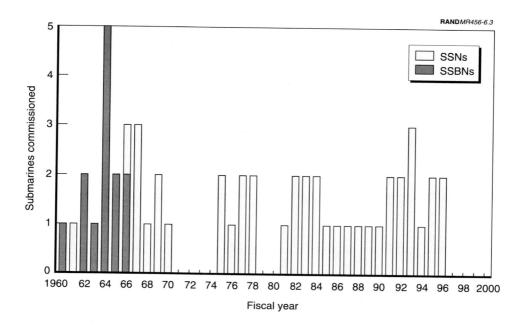

Figure 6.3—Newport News Submarine Production History

nor NSSN require refueling and thus are not candidates for early decommissioning.

— Whether to *extend the nominal life* of the last 31 Los Angeles–class submarines and the Seawolves from 30 to 35 years. Only those ships are considered, because the earlier ones would probably need to be refueled a second time if they were to operate beyond 30 years.[3]

These constraints can be imposed in any desired combination, and the model will determine the new submarine construction schedule that minimizes the future cost of procuring, operating, overhauling, and decommissioning the submarines.

Before we go on, let us clarify the way we handled transitions in the production rate. We specified a maximum production rate, that is, a maximum number of deliveries per year; there were two kinds of instances in which a rate lower than the maximum was used. First, we input a "ramping up" schedule that specified the number of ships per year a shipyard could produce in moving from no ships delivered to the maximum rate (represented by the rows in Table 6.1). These schedules were designed to reflect in a generic way the delivery schedules (starting from an inactive yard) that were produced by the workforce buildup model described in Chapter Three. Second, in subsequent years the fleet replacement model could choose to deliver fewer than the maximum number allowed (e.g., in a year in which a smaller-than-usual number of decommissionings was scheduled[4]). In either case, the number of deliveries from a particular shipyard could not be changed by more than one from year to year.

Table 6.1

Submarines Delivered per Year for Different Maximum Rates

Max. Rate	Year of 1st Delivery (X)	X + 1	X + 2	X + 3	X + 4	X + 5
1 per year	1	0	1	1	1	1
2 per year	1	0	1	2	2	2
3 per year	1	0	1	2	2	3

NOTE: These are rates *per shipyard;* for cases involving two shipyards, the combined initial schedule would be 2, 0, 2, ...

---

[3]We choose a five-year extension because it is large enough to make a difference in cost and schedule but not so large as to merit dismissal out of hand on feasibility grounds. As explained in Chapter Seven, however, extensive study will be required to determine the feasibility of *any* life extension plan.

[4]The number of decommissionings per year is also partly under the model's control through the option to decommission certain vessels early.

Besides the constraints listed above, there are two other inputs to the model: the current inventory of attack submarines and cost data. The inventory is described in Chapter Two. The cost values used in the analysis are shown in Table 6.2[5] (for more detail on O&S costs, see Appendix G). While all such costs are subject to variation from boat to boat, and subject to the inevitable uncertainties of estimating future costs, such lack of precision is unlikely to have a significant effect on the final study results.

There are several additional points regarding our approach to costs. Most of the analysis performed in the study has been on the basis of minimizing the costs, at a 5 percent discount rate, over the time period of 1994 to 2030. Sensitivity runs with larger discount rates produced negligible differences in the calculated schedule of new ship production. The end date of 2030 was selected because by then all of the presently existing ships will have been retired (if the 30-year lifespan is retained; see Figure 6.4) and the entire quantity of new ships needed to sustain the designated fleet size will have been constructed.

In the calculation of cost distribution over time, we assume that the entire procurement cost of a new submarine is incurred at a single point, six years prior to commissioning. Other costs, such as overhaul, refueling, and decommissioning, are incurred in the year that the activity occurs. This treatment of costs reflects our interest in budgetary effects—appropriations as opposed to outlays.

Table 6.2

Values of Cost Inputs to the Fleet Replacement Model
(costs per ship, millions of 1992 dollars)

Cost Input	688-Class			Seawolf-Class	NSSN
	688–699	700–718	719–773		
Initial construction	800	800	800	1750	1000
Refuelings and overhauls[a] at					
7 years	175	90	90	200	175
16 years	265	265	175	200	175
24 years	175	175	265	200	175
Decommissioning	50	50	50	50	50
Annual operations	15	15	15	17	15

[a]$265M actions are refueling overhauls, $175M–$200M actions are regular overhauls, and $90M actions are depot modernization periods.

---

[5]The costs in Table 6.2 are based on information available to us in the spring of 1993. The Navy has recently informed us that it plans to overhaul Seawolf-class ships and NSSNs only once and, as noted above, that it does not plan to refuel the later 688s. (However, if ship life were to be extended, a refueling at 24 years would be likely.) The Seawolf and NSSN changes would have virtually no effect on our cost analysis, as the total overhaul cost would not be different (see Appendix G). The effect of the 688 change is not readily predictable.

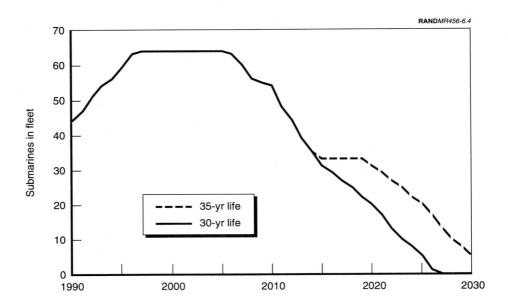

Figure 6.4—Los Angeles- and Seawolf-Class Submarines Will All Be Retired by 2030
(unless lifespan is extended)

We assume procurement costs are invariant with quantity. Experience suggests that the learning curve slope in serial production of a submarine class is typically rather flat, and is further flattened by the practice of introducing design improvements from time to time. In addition, given that we anticipate a smaller production rate than in the past and a more varied threat, fewer ships may be built between substantial design modifications. Thus, we believe that introducing an arbitrary learning curve would add nothing to the quality of the model results. Introducing a learning curve would also be inconsistent with our analytic strategy of making assumptions that favor longer gaps.[6]

Cost and schedule effects of reconstituting the industry after a gap in production have not been incorporated into the model. To simplify the analysis process, those effects are examined separately (as described in Chapters Three, Four, and Five) and combined with the model results during the final evaluation of policy alternatives in Chapter Seven.

---

[6]If submarines cost less with a longer gap because of discounting, applying the same percentage cost reduction to analogous ships in shorter- and longer-gap construction sequences will reduce the absolute cost difference between the two.

DEFINING AND SCREENING THE ALTERNATIVES

In analyzing shutdown and reconstitution, we ignored eventual fleet dynamics, began with the gap now under way, and estimated costs and delays for various restart years. Now, we begin at the end, with one of a range of fleet sizes that might be chosen as a goal, and work backward. At some point after the decommissioning of Los Angeles–class submarines begins, the fleet size will drop below the goal unless submarines have by then been produced in large enough numbers to make up the difference. Once values have been assigned to three factors—the target fleet size, a production rate, and a set of lifespans for current ships—the latest possible restart date (and thus a maximum gap length) can be determined. Limiting the values of the factors (or model constraints) to ranges of reasonable, practical choices limits the gap lengths achievable.

By combining the various possible constraint values listed above, we obtained 80 alternative fleet replacement strategies for further consideration. In doing so, we had to limit our treatment of shipyard-and-production-rate combinations or deal with even more strategies. We chose to consider all three single-shipyard cases (for maximum rates of one, two, and three per year). However, we confined ourselves to two two-shipyard cases: one and two per year and two and two per year. The former gave us a case in which a sustained rate of three per year could be achieved more rapidly after restart than with a single shipyard, and the latter furnished a combined four-per-year rate. It did not seem to us that cases of one and one or one and three per year would add anything to what we already had, and combined rates of more than four per year appeared unnecessary.

The 80 alternative strategies are shown in Figure 6.5. Each triangular cell represents an alternative. For example, the lightly shaded triangle represents a fleet replacement strategy in which an eventual fleet size of 30 is reached through a maximum construction rate of three ships per year at a single shipyard; the maximum current ship life is 35 years and early decommissioning is permitted. In the strategy represented by the darkly shaded triangle (partly hidden), the first three factors are the same, but the maximum age is the current 30 years and no ship is decommissioned early. This kind of diagram is useful for displaying model outputs, such as cost, in the cells. We use it for year of first delivery or restart in repeating portions of this figure below.

Narrowing the Range of Production Rates Considered

It was possible to screen out some alternatives through consideration of schedule issues alone (without estimating costs). For example, a production rate of one ship per year is obviously not enough to sustain, when steady-state conditions prevail, a fleet of more than 30 or 35 ships if those ships must retire

RAND*MR456-6.5*

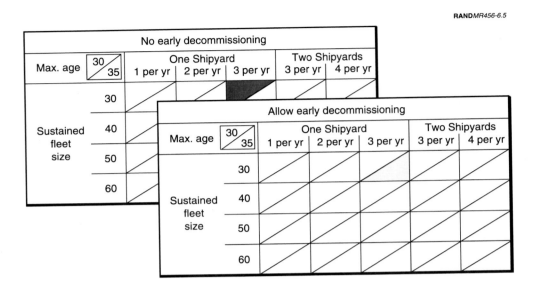

Figure 6.5—The 80 Alternative Fleet Replacement Strategies Considered

after 30 or 35 years. Can it sustain even 30 when steady-state conditions do not prevail (i.e., when ships are coming out of the force at a rate of three per year, but when the starting point is more than 60 ships)? Let's examine a fleet replacement strategy that places minimum stress on future production resources: We assume the Seawolf-class SSN-23 is built, production of the NSSN is started relatively early with first delivery in 2005, and no ships are decommissioned early (so they need not be replaced early). A simple spreadsheet analysis produces the fleet profiles in Figures 6.6 (maximum age of 30 years) and 6.7 (maximum age of 35). If the current 30-year retirement age is retained, the fleet size will drop below 25 if only one ship can be delivered per year. Even if the later Los Angeles–class ships are not decommissioned until age 35, the fleet will still drop below 30. Thus, a maximum production rate of one per year will probably not be adequate to meet future fleet needs.

This analysis permits a second useful conclusion. A production rate of three ships per year at two shipyards will sustain a fleet size of 60, even with all ships decommissioned at age 30. Thus, rates greater than three per year will probably not be needed in the future. Elimination of the one-per-year and four-per-year production rates allows deletion of the first and fifth of the five columns in Figure 6.5.

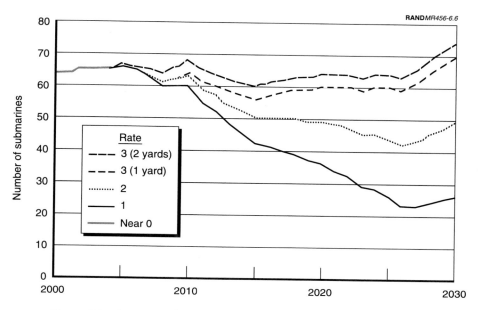

**Figure 6.6—Largest Achievable Fleet Size When Maximum Ship Age
Is 30 Years**

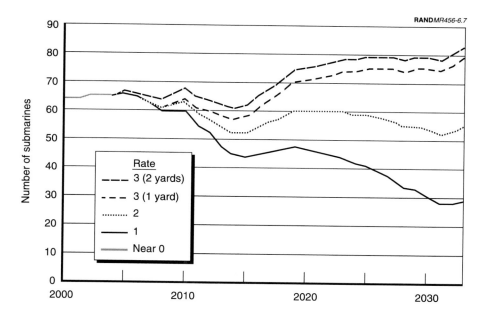

**Figure 6.7—Largest Achievable Fleet Size When Maximum Ship Age
Is 35 Years**

A third valuable inference can be drawn from a comparison of Figures 6.6 and 6.7. For each production rate, extending the life of the later 688s allows a larger fleet size to be maintained. The gain is small for three ships per year and is not enough for one per year, but at the median rate of two per year, life extension permits a substantial gain (25 percent) across a potentially important range of fleet sizes. We will have more to say about the value—and drawbacks—of life extension in the next chapter.

Choosing One of the Two Decommissioning Strategies

The different fleet replacement strategies obviously have different consequences for both near-term and far-term costs. Increasing the allowable production rate, for example, lowers present discounted costs by allowing production to be put off further into the future; however, it does not affect the sum of then-year, undiscounted costs. Both discounted and undiscounted costs can be saved by decommissioning some of the Los Angeles–class submarines early instead of refueling and continuing to operate them. This suggested a second screen, so we ran some cases through the fleet replacement model.

In Figure 6.8 we show the fleet profile consequences of two cases that differ only in the decommissioning rule applied. The top curve shows what happens if we do not allow early decommissioning of 688s, while for the second curve, early decommissioning is allowed. In the latter case the model decommissioned 24 submarines early, at their nominal refueling point. (That was the number minimizing cost given the parameters shown; for different parameters, different numbers would be decommissioned.)

For the pair of cases shown, early decommissioning permits a reduction in total undiscounted costs of about $14 billion (in 1993 dollars; see Figure 6.9). This is a consequence of saving $440 million per boat by eliminating refueling and overhaul costs ($265 million for the more recent 688s[7]) plus $210 million in operations costs for the forgone operational period ($90 million for the more recent boats).

These cases illustrate a cost relationship that turned out to be true in every situation, although of course the amount of savings varied from one case to another. As a consequence, in all subsequent analysis we invoke early decommissioning of the 688s to the extent necessary to minimize cost while still

---

[7]The maintenance schedule for the first 31 Los Angeles-class ships calls for an overhaul following refueling, while that for the second 31 does not. The advantages of decommissioning apply principally to the first 31 ships. Decommissioning the later ships early (at age 24) saves less money, and it forgoes savings from life extension to 35 years and from postponing production of replacement ships (see Chapter Seven).

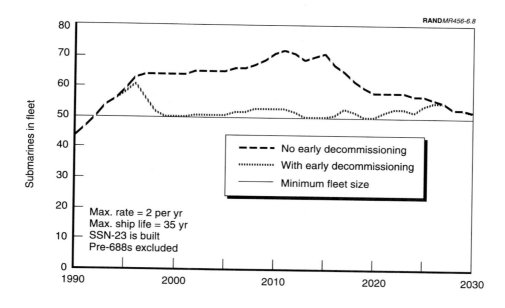

Figure 6.8—Target Fleet Size Can Still Be Sustained If Some Submarines Are Decommissioned Early

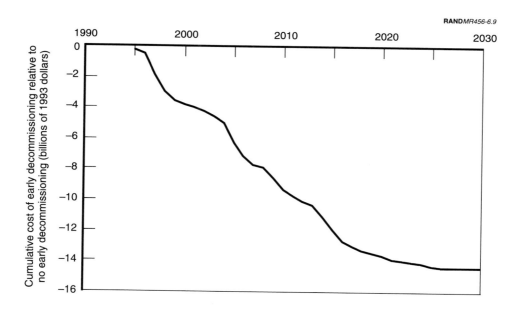

Figure 6.9—Early Decommissioning for the Case Shown in Figure 6.8 Would Save $14 Billion (undiscounted)

meeting force size goals and not exceeding maximum new construction rates. That makes the back panel in Figure 6.5 unnecessary.

Narrowing the Range of Fleet Sizes Considered

Some might argue that the range of fleet sizes we consider is already too narrow; perhaps in the security environment of the early 21st century the United States will need only 20 attack submarines or fewer. In that case, it would neither be necessary to deliver the next submarine until 2020 (see Figure 6.4) nor to reconstitute the submarine production base until after 2005. This could lead to different conclusions regarding the merits of extending the production gap.

Evaluating the merits of such a position is outside the scope of this analysis.[8] We draw the range of fleet sizes we consider from those most widely discussed by knowledgeable participants and observers of submarine fleet requirements and acquisition. Even a 30-ship fleet is outside that range. We retain it for cursory analysis of a single alternative (maximum rate of two per year, maximum ship life of 30 years). With that one exception, we eliminate the top row in Figure 6.5.

Checking the Feasibility of Production Gaps

The final step in the preliminary screening of future fleet replacement options was to examine the maximum possible gap that might be sustained for each of the various cases.   That is, for each of the remaining cases, how long can delivery of the first ship and thus start of production be put off? The delivery and restart dates for each of the remaining cases are shown in Figures 6.10 and 6.11, respectively. Because we are interested in both the costs and benefits of a sustained production hiatus, we do not interrupt the gap with an SSN-23.

Four of the cells in Figures 6.10 and 6.11 are blank. For these alternatives, delivery of the first ship would have to be taken before 2004, implying a restart before 1998. For that to happen, the design process for the NSSN would have to be further along than it is now. At a maximum rate of two ships per year, production cannot be restarted soon enough to prevent the fleet size from dropping below 60, with or without life extension to 35 years. Nor can it be restarted

---

[8]It may be worth noting, however, that a 20-ship fleet would leave the United States with many fewer attack submarines than Russia and China each now has and with about the same number that North Korea has.

RAND*MR456-6.10*

Allow early decommissioning			
Max. age 30 / 35	One Shipyard		Two Shipyards
	2 per yr	3 per yr	3 per yr
Sustained fleet size 40	2005 / 2010	2010 / 2010	2011 / 2011
50	/ 2005	2006 / 2007	2008 / 2008
60	/	/ 2004	2005 (59) / 2005

NOTE: No SSN-23; blank triangle indicates delivery needed earlier than is feasible.

Figure 6.10—Latest First-Delivery Date Feasible for Each Production Strategy

RAND*MR456-6.11*

Allow early decommissioning			
Max. age 30 / 35	One Shipyard		Two Shipyards
	2 per yr	3 per yr	3 per yr
Sustained fleet size 40	1999 / 2001	2001 / 2001	2001 / 2001
50	/ 1999	1999 / 2000	2000 / 2000
60	/	/ 1998	1999 (59) / 1999

NOTE: No SSN-23; blank triangle indicates restart needed earlier than is feasible.

Figure 6.11—Latest Restart Date Feasible for Each Production Strategy

soon enough to sustain a fleet size of 50 if the current service life is retained. The same goes for 60 at three ships per year. These four cases are thus dropped from further consideration.[9]

For the remaining strategies, initial deliveries can be taken anywhere from 2004, for the one remaining case in which a single shipyard must sustain a 60-ship fleet, up to 2011, when two shipyards work to sustain 40 ships. However, this seven-year delivery differential shrinks to three years at restart if the yards are inactive after current production ends. That is because current production ends around the time of restart for a 2004 delivery, and many workers would be immediately available for the new production line. But if restart is postponed only a few years, those workers will be gone, so it will take much longer to build up the workforce and much longer to build the first submarine (see Figures 3.10 and 3.11).

Thus, gap options are limited in practice to a range of choices spanning only three years. That limits the savings that can be achieved from deferring restart as long as possible, as we shall see in the next chapter.

Before continuing, we make a point about fleet replacement dynamics. We have been showing maximum feasible gaps to give extended-gap strategies a "fair shake" by maximizing their discounted savings. However, such gaps would impose a serious penalty on future generations. Consider, for example, the case of sustaining a 40-ship fleet with a production rate of up to three submarines per year. That would allow authorization of the first new submarine to be delayed until 2001, with delivery in 2010. Production would then proceed for 17 years, with the 40th submarine being delivered in 2027 just as the last of the present submarines reaches the current retirement age. However, the Navy in 2027 would find itself with a fleet of 40 submarines, the oldest of which would be 17 years old. Continuing the presumption of a 30-year life, no new submarines would be needed for another 13 years. The cycle of feast and famine would repeat every 30 years. In fact, at least a limited problem of this type is built into any policy that delays new construction as long as possible. A forward-looking planner would probably want to produce at a rate close to that needed to sustain the anticipated future fleet size in steady state, which would require a relatively early start date.

---

[9]A special case exists for a fleet size of 60, maximum age of 30 years, and production rate of three per year from two shipyards. This combination leads to a force size decaying to 59 ships before starting to build back up. That is sufficiently close to 60 ships that we elected to retain this case in the analysis.

Chapter Seven

COMPARING THE COST OF ALTERNATIVE STRATEGIES

We next compare the costs of fleet replacement or "gapping" strategies that survived the previous chapter's preliminary screening. We combine the costs from the fleet replacement model, which takes into account construction, maintenance, and operations, with the reconstitution costs estimated in Chapters Three, Four, and Five.[1]

The strategies surviving the preliminary screening are those represented by the dated triangles in Figures 6.10 and 6.11. We limit our analysis to strategies intended to sustain a fleet of 40 or 50 ships from a single yard (plus a single 30-ship comparison). The two-shipyard cases offer only minor extensions of the production gap. The sole strategy for maintaining a 60-ship fleet from a single yard requires both a high production rate and an extended ship life, and the first ship would still have to be delivered a year earlier than in any of the other strategies. It would thus not appear that this strategy is a likely choice.

The dated strategies in Figures 6.10 and 6.11 are "maximum-gap" strategies: no more submarines will be started until the one represented by the restart and delivery years shown. For each of these, we define a corresponding "minimum-gap" strategy. The minimum-gap strategies are intended to be the closest we can now come to "continued production." These entail beginning SSN-23 in 1996 for delivery in 2002, and beginning the first NSSN in 1998 or 1999 for delivery in 2005. Except for the restart date and the inclusion of SSN-23, the characteristics of each minimum-gap strategy are the same as those for the maximum-gap strategy it is paired with. As an example, we show in Figure 7.1 the way that gapping and restarting are related for one pair of minimum- and maximum-gap strategies (minimum fleet size of 40 ships, maximum production rate of two per year, and maximum ship life of 35 years).

---

[1]The cost of reconstituting the nuclear-core vendor is not included, as we do not anticipate that vendor will have to shut down.

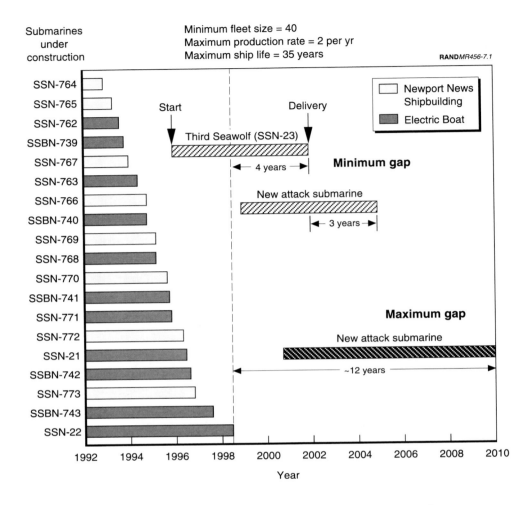

Figure 7.1—Gap and Restart Relations Between Minimum- and Maximum-Gap Strategies

Given the specified restart date (and thus the gap length), the fleet replacement model determines the schedule of construction, early decommissioning, and life extension (where permitted) that will replace the current fleet for the lowest net present value. The schedule must meet the requirements of maximum production rate and minimum fleet size that characterize the strategy being assessed, and must avoid further production gaps by delivering at least one ship per year.[2] After reconstitution costs are added to the costs of the optimal

---

[2]With certain early exceptions. No ships are built between SSN-23 and the first NSSN, and, as noted in Chapter Six, we allow for the time needed to reconstitute the workforce by assuming the second ship is not delivered until two years after the first.

schedule identified by the model, we can compare the net present value of the total costs with that for a schedule derived from a different restart date.

Note that we are comparing cost-minimizing schedules *contingent on restart dates* defined outside the model. We do not attempt to discover the restart date that minimizes costs. From limited analyses along those lines, we believe that, at least in some cases, the most economical restart strategy may fall *between* the two strategies we do compare—the minimum and maximum length gaps. But, given the small gap range feasible, we prefer to compare schedules that differ as much as possible, under the constraints applicable.

As in Chapter Three, we analyze in detail one illustrative set of alternatives and summarize the results of the other assessments. For the illustration, we have chosen a fleet size of 40 at a maximum production rate of two per year. We compare minimum and maximum gaps, and 30- and 35-year maximum ship lives. Results for three per year and for the 50-ship cases are given in this chapter only in broad outline. Appendix H provides a fuller treatment.

SUSTAINING A FLEET SIZE OF 40 SHIPS AT TWO PRODUCED PER YEAR

Differences in Discounted Costs

In Figure 7.2, we show the cumulative cost of sustaining the attack submarine fleet (688s, Seawolves, and NSSNs) from 1994 to 2030. By "sustaining," we mean the costs of reconstituting the production capability after the current gap, building ships, maintaining them through scheduled overhauls and refueling, operating the boats at current tempos, and decommissioning them. We count all these costs because all are affected by the choice of fleet replacement strategy. The length of the gap influences reconstitution costs. Decisions regarding early decommissioning and life extension affect the amount of refueling and overhauling done. All these choices interact to determine the number of ships that need to be built and thus eventually decommissioned. They also determine the number of ships in the fleet at any given time, which affects the cost of operations. Furthermore, the timing of all these costs affects their present value. Costs are discounted to the beginning of the period at 5 percent per year.

From the curves in Figure 7.2, we draw two conclusions, which, as we will show, are applicable in all our analyses, regardless of fleet size and production rate:

- The cost differences among the various strategies are modest relative to the total costs of sustaining the fleet and to the uncertainty of our prediction methods.

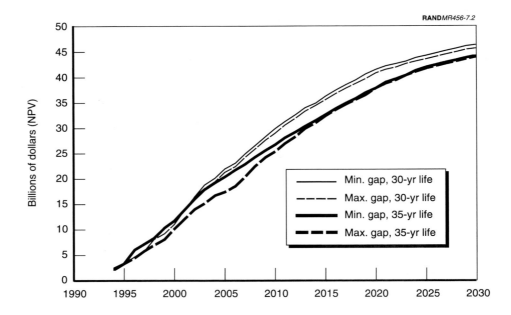

Figure 7.2—Cumulative Cost of Sustaining the Attack Submarine Fleet (40-Ship
Minimum) at a Maximum Production Rate of Two per Year, Discounted at
5 Percent per Year

- The cost advantage of extending ship life from 30 to 35 years is greater than
  the advantage of extending the production gap (see the blow-up in Figure
  7.3).

Provisos should be attached to each of these. We use the sum of all fleet-related
costs as the basis for comparison because those are the costs affected by the
choice of replacement strategy. However, it could be argued that our denomi-
nator is too large—and our estimate of relative costs thus too small. It could be
argued that one does not choose between fleet replacement schedules with the
objective or even the expectation of substantially affecting (for example) the
cost of operations. Indeed, the strategies differ in operational costs by less than
4 percent. It could also be argued that the $2.4 billion difference between the
most and least costly strategies is a lot of money, regardless of the basis of comparison.

The principal reason we choose to be cautious in asserting cost differences is
not because they are small relative to total costs (apparent as that may be from
Figure 7.2), but because they are not large enough relative to our possible estimating error. We have already discussed the uncertainties attending reconstitution cost estimates, and our estimate of the cost of building the NSSN could

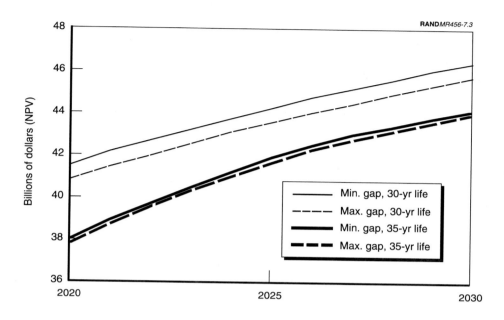

Figure 7.3—Upper Right Portion of Figure 7.2 at High Resolution

also prove conservative. We cannot be confident that costs estimated within a billion dollars or so of each other actually vary in the direction shown.

As for our second conclusion, the only additional cost we take into account in assessing life extension is that of an overhaul assumed required when a ship reaches 30 years of age. Other costs, however, should be anticipated. The U.S. Navy has essentially no experience in operating submarines beyond the standard 30-year life (two have been operated for 31 years). It may be necessary to inspect, monitor, and test hull elements and critical components on ships approaching the 30-year mark to discover trends that can be extrapolated. If it is discovered that certain components have too high a probability of failure between 30 and 35 years, those components would have to be replaced at the 30-year overhaul, increasing costs. It is not inconceivable that the total unaccounted cost of extending submarine life from 30 to 35 years will exceed the NPV savings shown in Figure 7.2. It is also possible that it will prove infeasible to extend service life. However, given the potential savings, a serious look at this option would seem advisable.

Differences in Undiscounted Costs

In Figures 7.4 and 7.5, we show the results from Figures 7.2 and 7.3 in undiscounted terms. Again, the differences are small relative to total costs. We show

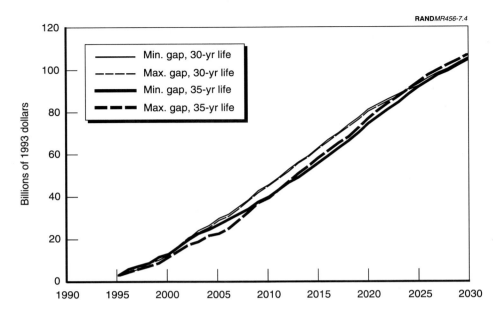

Figure 7.4—Cumulative Cost of Sustaining the Attack Submarine Fleet (40-Ship Minimum) at a Maximum Production Rate of Two per Year, Undiscounted

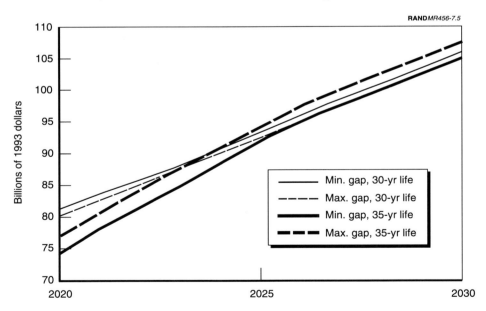

Figure 7.5—Upper Right Portion of Figure 7.4 at High Resolution

these results because they indicate the effects of discounting, and because we intend to use undiscounted costs below to analyze the reasons for differences among the strategies. Caution is urged in their interpretation and use, for two reasons:

- Decisions regarding spending and investment over the long term should be made on the basis of discounted costs. Some would say that no saving results from postponing a billion-dollar purchase by ten years, that it's "just moving money around." But that ignores the value most people would ascribe to having the benefits of that purchase now rather than later (whether that money goes to a submarine or some other purpose).

- The costs we show in Figure 7.4 are the result of converting the discounted output of the fleet replacement model into undiscounted terms. The output of the model is the result of fleet replacement decisions made to minimize *discounted* costs. If submarine acquisition policy is to be made on the basis of undiscounted costs, it should be based on fleet replacement decisions that minimize undiscounted costs.

Identifying the Sources of the Cost Differences

Notwithstanding the cautions we have urged in interpreting predicted cost differences among the strategies, we believe these differences merit further examination, for two reasons. First, the larger differences—those on the order of $2 billion or more—may represent genuine differences in the directions shown, even if assumptions regarding reconstitution and construction costs are changed. We recognize that long-term policy decisions must often be based on uncertain data, that some information is better than none, and that policymakers might not wish to ignore predicted cost differences of that magnitude. The reasons for these differences may thus be of interest for policymaking.

Second, an analysis of the sources of the cost differences—even the smaller ones—is useful because it demonstrates how the need to meet certain objectives and constraints affects not only construction but also maintenance and decommissioning decisions. The decisions that need to be made to minimize cost are not always the ones that would have been expected without running the model. The results thus illustrate the utility of analytic modeling of this type in planning submarine acquisition over the long term.

We begin with a different way to show the costs in Figures 7.2 and 7.4. Figures 7.6 and 7.7 display the same discounted and undiscounted profiles of future

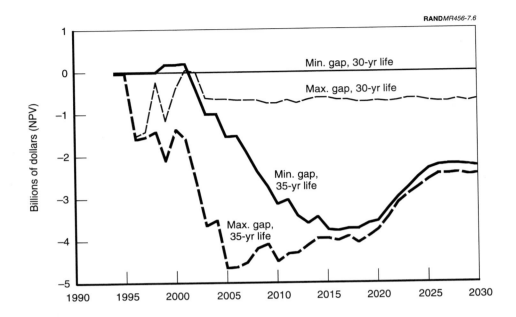

Figure 7.6—Cumulative Cost of 40-Ship, Two-per-Year Strategies, Relative to
Cumulative Cost for Min.-Gap, 30-Year-Life Strategy, Discounted

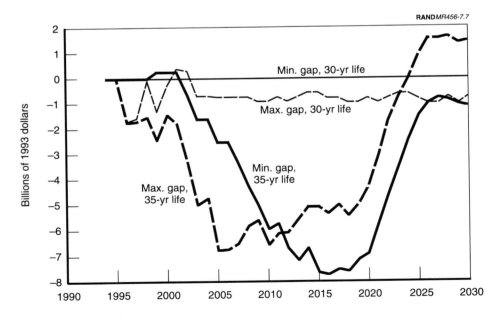

Figure 7.7—Cumulative Cost of 40-Ship, Two-per-Year Strategies, Relative to
Cumulative Cost for Min.-Gap, 30-Year-Life Strategy, Undiscounted

cumulative costs, this time in terms of their difference from the 30-year-life, minimum-gap case. The latter thus becomes a flat line at zero. It is clearer on these curves than on the previous ones that the cumulative difference in cost between strategies varies substantially over time.

We will now compare the minimum vs. the maximum gap, first assuming a 30-year maximum ship life, then 35 years. Then we will compare a 30-year life vs. 35 years, first assuming a minimum gap, then a maximum. We will concentrate on the undiscounted curves, as it is easier to discover the sources of cost differences when not simultaneously accounting for the time value of money. The differences among the discounted curves can then be readily understood.

Minimum vs. Maximum Gap: the 30-Year-Life Cases

These cases are represented by the two thin lines, solid for minimum gap (the horizontal line at zero dollars), and dashed for maximum, in Figures 7.6 and 7.7. If it proves impractical to extend the life of submarines to 35 years, this is the most interesting comparison. It also lends itself the least to support a decision one way or the other, as the cost difference between the two maximum-gap options is less than or equal to the undiscounted value of a single submarine over the entire time span of replacing the current fleet. In fact, the early swings in relative cost result from the purchase of single ships. The authorization of SSN-23 in the minimum-gap case gives a cost advantage to a continued gap, most of which is lost at the maximum gap's NSSN start in 1998, only to be returned at the minimum gap's NSSN start in 1999, and so on.

This situation is shown more clearly in Figure 7.8. We still show the difference between minimum- and maximum-gap strategies, but here the cost for each year is shown separately rather than cumulatively, and it is broken down by cost element. If, for a particular year, the minimum case requires, say, a higher construction cost than the maximum, that difference is shown as a black bar extending below the zero line. Conversely, if the maximum gap requires extra refueling, a dark gray cross-hatched bar extends above the line.[3] The early cost swings arising from the timing of ship construction show up clearly in this display. Eventually, the same number of ships is built in both cases. However, a cost difference of $760 million (in 1993 dollars) accumulates, almost all of it by 2003 (see Figure 7.7). Where does it come from?

The cost difference between the minimum- and maximum-gap cases can be viewed as arising entirely from the decision in the former case to bridge the gap

---

[3]In this and similar graphs for the following comparisons, we do not show the differences for decommissioning and operating costs, as these are small on an annual basis.

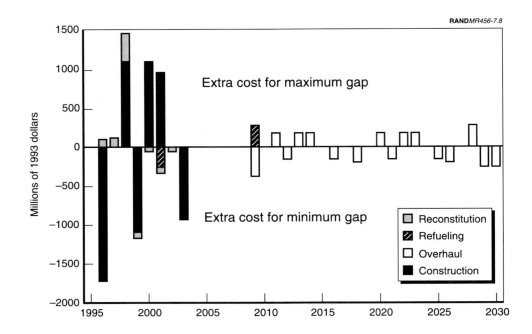

Figure 7.8—Annual Cost Differences, Maximum Minus Minimum Gap, Max. Ship Life 30 Years, Min. Fleet Size 40, Max. Production Rate Two per Year

with SSN-23, which we assume will cost $880 million more to build, operate, and maintain than an NSSN. This source for the cost difference is not surprising, because the principal production schedule difference between the two cases is the presence or absence of the third Seawolf. Restart dates for the minimum and maximum gaps are very close, as restart cannot be delayed beyond the late 1990s if a 40-ship fleet is to be sustained at a production rate of no more than two per year. Other differences between the two cases offset each other. The maximum gap requires an extra $250 million in reconstitution costs, while non-Seawolf overhauls and operations come to $220 million more in the minimum-gap case.

Because the cost difference accrues over the next decade, it is much the same in the discounted comparison. The principal conclusion arising from either comparison, however, is that the difference is too small for cost to play a significant role in deciding between the two strategies. The outcome could be reversed if just one of the many decisions constituting the two strategies were changed; indeed, given the uncertainties involved in the analysis, it could be reversed if none were changed.

Minimum vs. Maximum Gap: the 35-Year-Life Cases

If submarine life can be extended to 35 years, the cost relations between the minimum and maximum gaps change dramatically, as illustrated by the two heavy lines (solid and dashed) in Figure 7.7 (regraphed in Figure 7.9 to clarify the relationship). From 1999 through 2008, the difference between maximum and minimum gaps is at least twice as great as in the 30-year case, with the span reaching $4.2 billion in 2005 and 2006. From 2013 through 2030, the difference is again twice as great, but this time in the opposite direction, with a maximum spread of $2.9 billion in 2022. The outcome at the end of the period analyzed is a $2.6 billion cost advantage for the minimum gap (undiscounted).

The basic reason for the difference between this pair of curves and the previous pair lies in the relative gap lengths. In both cases, the minimum gap is the same: we assume that the earliest an NSSN can be delivered is 2005. As explained above, if submarine life cannot be extended beyond 30 years, that is also the *latest* the NSSN can be delivered and still sustain a 40-ship fleet at a production rate of no more than two per year. But if the service life of the later 688s can be extended to 35 years, the maximum delivery gap stretches to 2010. As a result, in the 35-year case the timing of ship construction differs for the two gap strategies, and that is reflected in the cost difference. From 1999 to 2003,

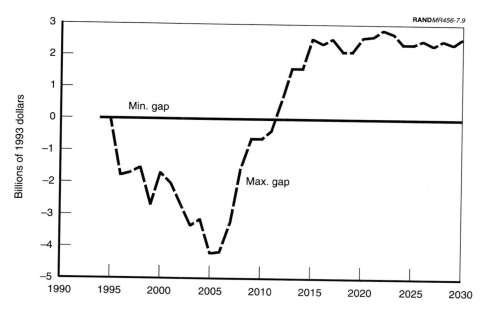

Figure 7.9—Cumulative Cost of Maximum Gap Relative to Minimum Gap, Max. Ship Life 35 Years, Min. Fleet Size 40, Max. Production Rate Two per Year, Undiscounted

about one ship per year is authorized under the minimum-gap strategy while the maximum gap has not yet ended (see Figure 7.10). From 2007 through 2013, one ship more per year is started under the maximum-gap strategy, and the maximum gap's cost advantage is eaten away. As in the previous comparison, the same number of ships is eventually built under both strategies. And, in fact, if ship construction alone were considered, the maximum gap would again come out ahead in savings, this time by $1.6 billion. Part of that results from (again) the substitution of a Seawolf-class submarine for an NSSN in the minimum-gap strategy.[4] Most of it, however, comes from a second factor: Delaying the maximum-gap restart date permits a more concentrated production schedule (fewer years in which production must be constrained to a single ship). We estimate a $75 million unit cost advantage to building two ships in a year instead of one,[5] and savings thus accrue to the maximum-gap strategy.

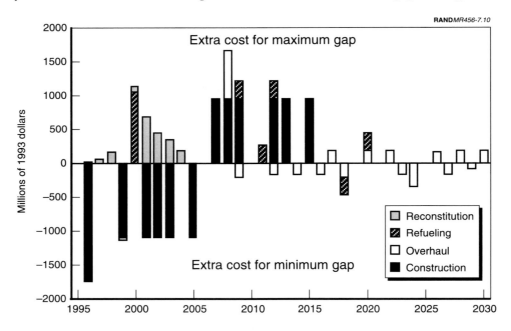

Figure 7.10—Annual Cost Difference, Max. Gap Minus Min. Gap, Max. Ship Life 35 Years, Min. Fleet Size 40, Max. Production Rate Two per Year

---

[4]For all of the minimum/maximum-gap comparisons discussed in this report, an intermediate case can be created that would be roughly equivalent to the minimum gap minus SSN-23. (For the first comparison, above, this "intermediate" case is actually the maximum gap.) The intermediate case would save on the order of $700 million (discounted or undiscounted) relative to the minimum gap.

[5]This results from allocating the $150 million fixed overhead cost to two ships instead of one.

If the maximum-gap strategy is substantially less expensive in terms of ship construction, what is responsible for the minimum gap's even more substantial cost advantage overall? First, although we constrain the minimum gap strategy to an earlier restart and thus a slower, costlier production schedule, that earlier start allows this strategy to take greater advantage of a significant source of cost savings: early decommissioning. Under the minimum-gap strategy, four more ships are decommissioned in 2000, resulting in $1.1 billion in refueling costs saved (plus further savings from overhauls and operations avoided).

Just as important is the difference in reconstitution costs. The late start under the maximum gap requires $1.9 billion more for workforce buildup, vendor reconstitution, and other costs than is needed with the minimum gap. The effect of the difference in refueling and reconstitution costs (the latter in particular) is to reduce the savings realized under the maximum-gap strategy on the left side of Figure 7.7 to the level shown. As a result, that level is not enough to compensate for the savings lost when the minimum gap has the construction advantage (i.e., fewer boats built and lower costs).

The pattern of savings could result in different strategy choices, depending on how the decisionmaker regards the time value of money. As shown in Figure 7.6, there is essentially no difference between the two strategies in the present value of costs discounted at 5 percent over the period through 2030, because the gains made by the minimum-gap strategy are later and thus discounted more heavily. A decisionmaker who took that approach to valuing future dollars would thus choose between strategies on other than economic grounds. But if he or she did not wish to discount future dollars, cost considerations would argue in favor of the minimum-gap strategy.[6] Finally, one who discounted future costs heavily, who was concerned almost exclusively about the next eight to ten years, would be attracted to the cost savings achieved by the maximum gap over that period.

30-Year Ship Life vs. 35-Year Life: the Minimum-Gap Cases

What is the effect of extending ship life and how does it compare with that of extending the production gap? Now we compare the two solid minimum-gap curves in Figure 7.7—the thin one representing 30 years and the heavier one 35.

In the 30-year case, high-rate production runs from approximately 2002 to 2020; in the 35-year case, from about 2014 to 2025. It is easy to see in the heavy curve in Figure 7.7 and in the bars in Figure 7.11 the resulting savings gain in

---

[6]We offer this as an illustration of how different discounting preferences could affect choices between gap strategies, but, as pointed out above, the model output is based on fleet replacement decisions that minimize *discounted* costs.

the 35-year case from 2002 to 2013 and in the 30-year case from 2021 to 2025. While we did not require that service life be extended in the 35-year case,[7] it is economical to do so for the typical boat. The gain from putting off production of the replacement boat is more than the costs we assume for life extension— that is, a refueling at 24 years[8] and an overhaul at 30 years.[9] Extending the lives of most (two-thirds) of the later 688s and Seawolves from 30 to 35 years permits postponing the point at which the entire current fleet must be replaced. The result is a $4.8 billion construction savings for the 35-year case. This is mostly offset by additional costs from two sources.

The production postponement in the 35-year case is achieved by refueling 688s at 24 years and keeping them in the fleet, thus putting off their replacement. These refueling costs come to $1.9 billion and fall between 2009 and 2014, slowing the savings gain for the 35-year case during that period (Figure 7.7). If refueling is advantageous, why is it not also done in the 30-year case? Granted, without life extension, refueling postpones ship replacement by only six years, but that still results in net discounted savings after figuring in the refueling cost.[10] The answer lies in the timing of refueling and the upper limit on annual production: Extra refuelings in the 35-year case are accomplished before high-rate production for the 35-year case but during high-rate production for the 30-year case. For a ship refueled at 24 years, construction must simultaneously start on its successor (for the 30-year case). This cannot be done when production is already at two per year, so the ship must be decommissioned rather than refueled. This is one way in which limitations on production rate can limit savings.[11]

---

[7]The last 31 Los Angeles–class ships were either refueled at 24 years and then overhauled for extension at 30 or were decommissioned at 24. We did not allow refueling at 24 and retirement at 30 in the 35-year case.

[8]There is little or no extra operating cost if a ship is refueled at age 24. For all strategies examined, the fleet size falls to 40 in 2008 and is held close to that level by a combination of decommissioning and new production through at least 2027. Adding or subtracting a ship by means of early decommissioning or life extension during that period is unlikely to affect operating costs, as another will be subtracted or added to hold the fleet size around 40. (Falling below 40 is not permitted and rising above it is uneconomical in most situations.)

[9]For an extension of *three* years, an overhaul would not be required at year 30, and almost as much might be saved as with a five-year extension, but at less cost (not quantified here) associated with ameliorating extension-related technical and safety risks.

[10]The choice here is between refueling and decommissioning at 24 years. Refueling at 16 years (instead of early decommissioning) entails the costs of a refueling, an overhaul, and 14 years of operations. Because the early 688s all reach the 16-year point by 2001, when the fleet size is well above 40, replacement is not required right away even if a ship is decommissioned early. The discounted savings from postponing production by refueling are thus small relative to the costs entailed, so refueling at 16 years is usually not advantageous.

[11]As noted in Chapter Six, we have recently learned that the Navy does not plan to refuel ships at 24 years in the 30-year-life case. The cores may permit service beyond that point; where they do not, the ships will be retired. We retain the discussion to point out that, in some scenarios, production

The other additional cost that contributes to offsetting the 35-year strategy's construction advantage is from overhauls. As the later 688s hit their 30-year marks between 2015 and 2027, they are overhauled for life extension. At the same time, NSSNs built before 2017 (mostly for the 30-year-case) are also coming up for overhaul. The result is a net of about one extra overhaul per year for the 35-year case (see Figure 7.11). This is reflected in the upturn in the 35-year curve in Figure 7.7 prior to 2020 and in its particularly steep ascent once the production difference kicks in. Following the end of life-extension-related overhauls in 2027, the earliest NSSNs reach their second overhaul point. Again, more of these were built under the 30-year strategy, and a small savings gain accrues to the 35-year curve. The cost of overhauling boats for life extension amounts to $1.8 billion, net of extra overhauls in the 30-year case for boats built early.[12]

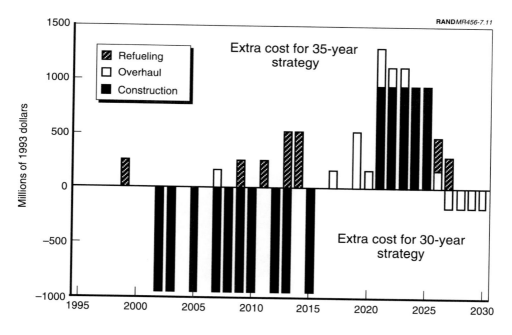

Figure 7.11—Annual Cost Differences, 35-Year Max. Ship Life Minus 30-Year, Min. Gap, Min. Fleet Size 40, Max. Production Rate Two per Year

---

rate limitations can limit savings and, more generally, that various scheduling parameters interact in complex ways to influence relative costs.

[12]All NSSNs, of course, will have the same overhaul schedule, but overhauls for those built later fall outside the window of analysis, as do more of the operating costs for those ships than for the ones built earlier under the 30-year case. The excluded costs are less than those accruing beyond 2030 for operating, overhauling, and decommissioning the extra boats built under the 30-year strategy.

Though the overall cost advantage of the 35-year strategy is small in undiscounted terms, it comes to $2.3 billion when discounted at 5 percent per year. Basically, the 35-year strategy allows fewer ships to be built[13] and large discounted savings by postponing construction of many that are built. This more than compensates for the cost of refueling more vessels and for the late, heavily discounted costs of extra overhauls at the 30-year point.

30-Year Life vs. 35-Year Life: the Maximum-Gap Cases

Here we compare the two dashed lines in Figure 7.7, the light one representing the 30-year case and the heavy one the 35-year case. The general relations between the curves and the reasons for them are similar to those for the minimum-gap cases, but here life extension runs up more undiscounted costs by 2030. The difference is in even greater refueling costs for the 35-year strategy in the current comparison, along with larger reconstitution costs (see Figure 7.12).

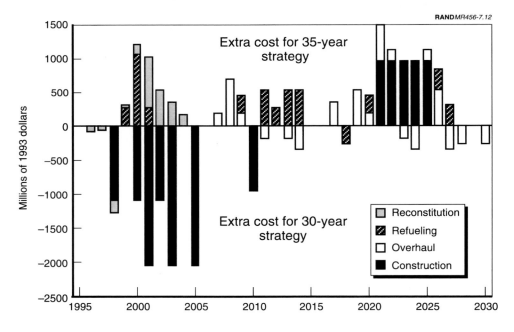

Figure 7.12—Annual Cost Differences, 35-Year Max. Ship Life Minus 30-Year, Max. Gap, Min. Fleet Size 40, Max. Production Rate Two per Year

---

[13]The fleet must be replaced five years earlier under the 30-year strategy, and then one ship per year must be built to avoid a production gap. Thus, at any point from 2008 to 2035 and beyond, more ships will have been built under the 30-year strategy. The extra ships cannot be "credited" against the next fleet replacement cycle to allow an eventual evening out of the numbers. The reason is that, over multiple cycles, it takes a slightly higher production rate to maintain a 40-ship fleet with a 30-year life than with a 35-year life.

Around the year 2000, the 35-year strategy refuels six more ships—instead of decommissioning them—than the 30-year stratregy. As pointed out above, decommissioning ships at age 16 (the age of these ships) is economically preferable to refueling them. But in the maximum-gap case, it is not possible to decommission these ships in the 35-year strategy because of the later construction restart. As it is, the fleet drops to 40 ships in 2008, and delivery of the first NSSN must wait until 2010 under the maximum-gap condition. Further early decommissioning would drop fleet size below 40. The later restart in the 35-year case also means greater reconstitution costs—about $1.7 billion greater.

The extra refueling and reconstitution costs lower the maximum savings accumulated in the 35-year case for this comparison relative to the previous one ($5.9 billion in 2005 vs. $7.6 billion in 2016). These are then outweighed by the construction and overhaul deficit that accumulates between 2015 and 2025. The extra refueling and reconstitution costs also keep the net present value of the 35-year strategy closer to that of the 30-year strategy than in the previous comparison. However, NPV still favors the 35-year case (by $1.8 billion) because of the heavy discounting applied to the late construction and overhaul deficit.

SENSITIVITY TO PRODUCTION RATE AND FLEET SIZE

In the following paragraphs, we show what happens to the relations among gap-length and ship-life strategies when we change production rate and fleet size. Results are summarized in the form of graphs analogous to Figure 7.6—cumulative, discounted cost relative to the 30-year minimum-gap strategy. For the 40- and 50-ship strategies, the full sets of graphs analogous to those presented above are given in Appendix H.

Sustaining a Fleet Size of 40 Ships at Three Ships Produced per Year

Raising the maximum production rate from two to three ships per year does not result in dramatic differences in total costs or in the cost relations among the gap-length and ship-life strategies (compare the end points in Figure 7.6 with those in Figure 7.13). This set of curves does vary from the earlier one in the course of expenditures over time. However, the pairwise comparisons follow the same general pattern: early construction-derived savings for the maximum gap or 35-year case, followed by loss of some of that advantage (in discounted terms) when high-rate construction begins under the postponed-production strategy. (As has been pointed out, the minimum-vs.-maximum, 30-year-life comparison at two per year is unusual because the maximum gap in that case is not really a postponed-production strategy.)

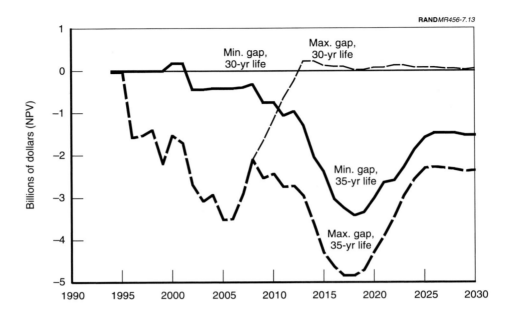

RAND*MR456-7.13*

Figure 7.13—Cumulative Cost of 40-Ship, Three-per-Year Strategies, Relative to
Cumulative Cost for Min.-Gap, 30-Year-Life Strategy, Discounted

Sustaining a Fleet Size of 50 Ships at Three Ships Produced per Year[14]

The total cost of sustaining a 50-ship fleet is, of course, more than that of sustaining a 40-ship fleet. However, the relations among the strategies are similar in the two cases (compare Figure 7.14 to Figure 7.13). Again, the costs of the strategies are all within a few billion dollars of each other, and changing maximum ship life has a bigger effect than changing the length of the production gap.

The biggest difference from the 40-ship case is that the maximum-gap strategies are no longer at parity or at an advantage with respect to the minimum gaps. The differences are not large, but they are consistent. For the larger fleet size, the maximum gap strategy with life extension is about a billion-and-a-half dollars worse off with respect to the corresponding minimum-gap strategy. It goes from having a small advantage for 40 ships to a small disadvantage for 50. For the 30-year case, the maximum-gap strategy is about $700 million worse off. Larger costs are run up relative to the minimum gap because production can-

---

[14]This chapter does not address the strategies to sustain a fleet size of 50 ships at a maximum rate of two per year. It is not possible to sustain such a fleet at such a rate with a maximum ship life of 30 years, so there is no baseline for comparison analogous to that in the other strategies discussed.

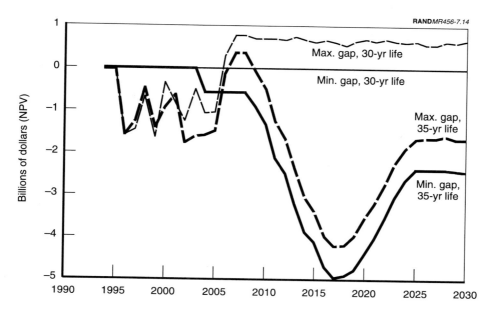

Figure 7.14—Cumulative Cost of 50-Ship, Three-per-Year Strategies, Relative to Cumulative Cost for Min.-Gap, 30-Year-Life Strategy, Discounted

not be put off as long when a 50-ship fleet must be replaced as it can when only 40 need be built.

Sustaining a Fleet of 30 Ships at Two Ships Produced per Year

As discussed in Chapter Six, a fleet size of 30 is below the level DoD regards as sufficient to meet U.S. security needs. However, since fleets of 30 or fewer may be proposed by some in debates over submarine funding, we compare minimum- and maximum-gap strategies for sustaining such a fleet, current service life assumed (see Figure 7.15).

Relative to the analogous comparison at 40 ships, two per year, the maximum-gap strategy saves more money with the smaller fleet size because production can be postponed longer (delivery need not be taken until 2010). The $2-billion short-term savings is at least as great as that for any other min.-max. comparison we analyzed for a 30-year service life. And, in contrast to the other comparisons, a modest savings—in excess of one and a half billion dollars—holds up over the long term.

Other analyses we conducted of fleet sizes below 40 exhibited a trend toward greater maximum-gap savings with smaller fleets. Savings from postponing

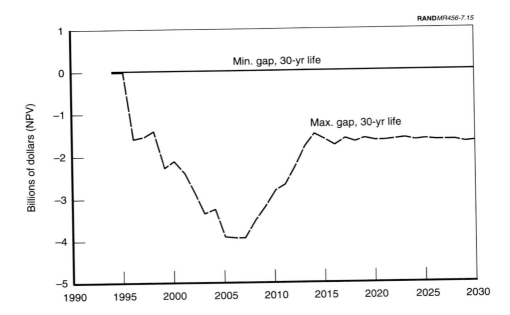

Figure 7.15—Cumulative Cost of Max. Gap Relative to Min. Gap for 30-Ship, Two-per-Year, 30-Year-Life Strategies, Discounted

production outweigh the greater reconstitution costs that postponement requires. However, as the ratio of fleet size to production rate drops, the risk of future production discontinuities increases (see the discussion at the end of Chapter Six). And the longer restart is postponed, the greater the unquantifiable program risk—a topic to which we now turn.

Chapter Eight

RISK

In Chapters Six and Seven, we quantified the cost and schedule effects of fleet replacement strategies involving production gaps of varying lengths. A principal conclusion of this analysis was that the net savings from an extended gap are small; in most comparisons of alternative strategies, they are so small that, given the uncertainty of our estimates, we cannot be sure they exist. When cost differences are small or uncertain, less quantifiable factors may play a particularly important role in decisionmaking. In planning submarine production, the key unquantifiable factors fall under the heading of "risk."

We discussed risk in Chapter Four in connection with nuclear-component suppliers, but we bring it up again here to emphasize its importance in the overall analysis and to address the topic somewhat more broadly. Uncertainties in the analysis are listed in Table 8.1, along with the manner in which they were treated in the *quantitative* analysis. More generally, extending the current gap in submarine production entails three kinds of risk:

- The risk that it will not be possible to reestablish the production program at affordable cost or on a satisfactory schedule.

- The risk that, during or after reconstitution, an accident will occur because of degraded construction quality.

- The risk that program benefits will be constrained or costs increased over the long term.

PROGRAM RISK

Taken by themselves, the cost and schedule estimates in the preceding chapters do not present a full picture of the costs and delays involved in extending the production gap. We made some allowance for the possibility of underestimates in our assessment of cost and time to reconstitute the nuclear vendor base, but

Table 8.1

Sources of Uncertainty and Their Treatment in the Quantitative Analysis

Source	How Treated
Willingness of shipyards to restart production	Assumed
Maintenance of engineering and design capabilities	Assumed
Maintenance of cadre during gap	Assumed, at DoD cost
Amount of fixed shipyard overhead	Fixed value used in main analysis; parameterized separately (App. F)
Early worker-attrition rate	Fixed value used in main analysis; parameterized separately (App. F)
Mentor:trainee ratio	Fixed value used in main analysis; parameterized separately (App. F)
Whether workers hired from other sub lines will have right skills for restart	Assumed
Percentage of laid-off workers rehirable	Fixed value used
Worker efficiency: pay ratios	Value is function of years of experience
Retention of management personnel	Assumed
Effect on public shipyards of redirecting work to sub yards	Assumed not to occur
Ability to keep nuclear core vendor in that business	Assumed
Willingness of other nuclear vendors to restart production	Assumed
Ability to reinvent nuclear technology and integrate new technologies after gap	Assumed
Nonnuclear vendor reconstitution cost	Value is function of gap length only
Industry support from non-sub-related sources	A few considered; generally, assumed to be limited
Acceptability of overseas suppliers	Assumed unacceptable
Cost of NSSN	Fixed value used
Minimum fleet size desired	Parameterized
Discount rate	Two values used
Maximum annual production rate	Two values used
Ability to extend ship life	Assumed
Learning curve for submarine construction	None assumed
Effect of future SSBN construction	None assumed
Source	How Treated
Pressures from interest groups against restart	Assumed absent
Granting of nuclear and environmental permits	Assumed, with no mitigation
Status of overall economy	Recent history (recessionary to low-growth) assumed

NOTE: "How Treated" refers to quantitative analysis only; variations from many of these assumptions are discussed qualitatively.

we have not done so elsewhere. Generally speaking, our estimates are based on the assumption that submarine production functions can be reconstituted without major impediments that would greatly increase the cost or time of doing so. But several such impediments loom.

We have allowed considerable cost and time to reconstitute technical and trade skills at the shipyards—up to $2 billion and 17 years to return to current capabilities. Even so, these may be underestimates. For example, our estimates are based on the optimistic assumption that the shipyards will be able to rehire 90 percent of workers laid off during the year preceding restart. But workers may leave the area or become disaffected in ways that reduce their propensity to work for the shipyard or compromise their effectiveness if they do so.

It may be difficult to sustain a large enough cadre of workers once the construction of current submarines ends. If the restart date is unknown or more promising prospects come along, workers may depart. It is critical that these workers be retained to serve as mentors at restart. While detailed drawings and specifications are prepared for each of a submarine's components and systems, we should recognize the contribution of experience. Many actions in submarine construction demand a particular experienced individual who is the only person in the yard or factory who can execute that process correctly or teach someone else how to do it. It may seem strange to some that the successful production of such a technologically advanced system could depend in part on a sort of "black art" or folklore, but that is indeed the case. There are many areas where the essence of production cannot be captured by a drawing, specification, book, or videotape. There is significant risk that, after a gap in production, the United States could not build another submarine to the same level of performance for some time—until a new base of experienced designers, technicians, and workers could be grown.

We have estimated the costs of rehiring management and support workers as a percentage of those for production workers. However, we have not assessed the specific contributions of management to an efficient production process. We estimated how production workforce drawdowns to various levels would influence reconstitution cost and schedule, but we did not do the same for management drawdowns. It is likely that released management personnel will be able to sell their talents to other industries elsewhere in the country and very unlikely that such people could be lured back.

Besides losing workers in an extended gap, the submarine industrial base will probably also lose firms—suppliers of various components used in submarine production. Companies "stung" once by a hiatus in submarine construction may be unwilling to invest in future construction. Future submarine construction may appear to be tenuous compared to the buildup in the 1960s. Companies may not care to accept the risk of reentering the submarine construction business, fearful that the U.S. government might again invoke another gap in production at some later time. Some companies may successfully make the conversion to civilian products and may not elect to re-convert to military applications.

We have only cursorily examined the effects of an extended gap on design and development expertise (Appendix A). Without some measure such as prototyping to fill the gap, it may be necessary to lay off experienced design engineers. Those who would otherwise have represented the next generation of recruits may seek more promising opportunities in the civilian sector. If both experienced designers and recruits are in short supply, there will be fewer opportunities to pass on the lessons learned in designing previous classes (some "black art" is involved here, too). Clearly, a deterioration in design expertise would affect the time required to reconstitute production, but the size of that effect is unknown.

While design capability may erode as the gap lengthens, the challenge to designers increases as technology progresses farther beyond the state of the art represented by the Seawolf class. Whereas the next submarine class need not meet, let alone exceed, some Seawolf performance criteria, it would be desirable to take advantage of those advances that *are* applicable to submarine operations in the reduced-threat environment. No one will want to ignore technological improvements that could afford U.S. submarine crews a more secure, efficient operating environment. Enhancements to the design process for the NSSN are intended to reduce technological risk. Some challenges will remain, however, not only for designers but for the production workforce. Our estimates of reconstitution delays based on workforce buildup do not take into account the possibility of extra time needed for testing or reworking improved systems.

Finally, restart could be delayed indefinitely by the need to satisfy regulatory requirements. Submarine construction yards and many component suppliers work with hazardous materials or processes that require licenses or permits from state or federal organizations (such as the Environmental Protection Agency, Department of Energy, and Nuclear Regulatory Commission). In many cases, the contractor's activities are "grandfathered" under previous regulations because of their continuing activity in that process, even though current regulations may be more strict. Contractors have legitimate concerns that their permits will lapse if they suspend activity. Regaining the permit or license may prove to be expensive and difficult, if not impossible, in a society in which environmental concerns, and nuclear ones in particular, often dominate. The regulatory process will provide a "foot in the door" to any who may wish to delay indefinitely or terminate the nation's nuclear submarine program.

Clearly, there may be many hurdles to clear in restarting submarine production. We cannot characterize well the costs and delays involved because no one has yet attempted to restart a dormant industry that produces technologically ad-

vanced products. As pointed out in Chapter Four, the few crude available analogues to restarting submarine production suggest that it may be as difficult
and costly to rebuild some capabilities as it was to build the current ones.

The extent of rebuilding required, however, depends on the length of the production gap. The gap has already been long enough to threaten the survival of
some nonnuclear vendors even if it lasts only until the scheduled SSN-23 start
in 1996. Other ill effects may not be realized unless restart is delayed into the
next decade. Thus, in attempting to balance qualitative program risks against
quantitative savings, greater weight should be ascribed to the former as gap
length increases.

ACCIDENT RISK

We discussed the risk of accident in some detail in Chapter Four. Suffice it to
say here that many of the factors mentioned above as contributing to program
risk also increase the possibility of an accident resulting in severe damage to or
loss of a ship during construction or operations. In addition to the immediate
physical consequences, such an accident, particularly a nuclear one, would
greatly raise the odds against successfully reestablishing the submarine construction program. Obviously, the Navy cannot rely on a "learning curve" or
trial and error in nuclear-submarine safety.

LONG-TERM RISK

A lengthy production gap will have negative consequences many years after it is
over. The longer the gap, the greater the risk that fleet sizes will not meet national security needs and the greater the risk that another production gap will
eventually result. Even if a submarine production program can be successfully
reestablished after a lengthy gap, acquisition planners will face an unpalatable
choice between low sustainable fleet sizes and high production rates. Or there
may be no choice—both may be necessary.

As shown in Chapter Six, even with a minimum-gap strategy the fleet size will
drop almost to 40 around 2025 if the production rate is held to two per year and
submarines are retired at 30 years of age. That minimum fleet size will drop if
the gap lengthens. If global tensions continue to abate, that may not be seen as
a major problem. But events of recent years have demonstrated the unpre-
dictability of the geostrategic environment. Developments in Russia, China, or
elsewhere may lead U.S. defense planners to conclude that the nation would be

safer with 50 attack submarines. That number cannot be achieved after a long gap unless at least three ships can be built per year.[1]

Building three ships per year creates its own long-term problems. It will sustain a fleet size of 90 at steady-state and current submarine life spans. If fewer than 90 ships are desired, steady-state cannot be achieved; the production rate must drop at some point. As shown in Chapter Six, a possible result is the rate will go to zero and another production gap will result, with all the risks described above. The risk of another gap can be lowered by initiating submarine production soon and building at a deliberate pace, instead of waiting a long time to undertake a crash program.

_____

[1]And, of course, it would take years to deliver the first submarines following production restart. Failure to hedge against future threats is another problem with the 20-ship minimum fleet size mentioned in Chapter Six. In that case, it could be cost-effective to wait 15 years or more between the start of the second Seawolf and the NSSN, but the ability to ramp up production quickly would be lost. Furthermore, the number of submarines available to counter a serious threat would be much smaller than the 40 taken as a minimum in this analysis.

Chapter Nine

CONCLUSIONS AND RECOMMENDATIONS

Our principal conclusions fall into three categories: those relating to the practicality of an extended production gap for attack submarines and those relating to the cost-effectiveness and the risks of such a gap. As these already emerge from the analyses of the preceding sections and in the appendixes, we simply restate them here to keep discussion to a minimum. For a more fully argued précis of the study, see the "Summary."

ON THE PRACTICALITY OF AN EXTENDED GAP

- Production schedule options are limited. Construction of the first submarine of the next class (termed the "New Attack Submarine," or NSSN) probably cannot be started before 1998. (A third submarine of the Seawolf class may be built in the interim.) But construction of the first NSSN *must* start by about 2001 if a fleet of submarines close to that now planned is to be sustained at reasonable production rates. The difference between the shortest gap now feasible and the longest practical is thus only three years.

- The longer the gap, the more difficult it will be to sustain a fleet adequately sized to the nation's security needs and the greater the risk that the concentrated production program then required will lead to another gap. If the next submarine is not started until after 1999 and ships are still retired at the age of 30, it will not be possible to sustain a fleet size of 50; a production rate of three per year would be required to keep the fleet from falling below 40 ships.

- If the more recently built Los Angeles-class submarines (the last 31) could be operated until age 35, greater flexibility in production scheduling could be realized. It would be possible to sustain a greater fleet size at the same production rate or the same fleet size at a lower production rate than would be the case with the current decommissioning age.

ON THE COST-EFFECTIVENESS OF AN EXTENDED GAP

- It is not clear that an extended production gap would result in any savings over the long term. For some combinations of desired fleet size and maximum production rate, savings may be realized by extending the gap; for others, losses may result. In all cases, the projected gains or losses are smaller than the errors that may accompany our prediction methods over that time frame, so they cannot be asserted with any confidence. However, it appears that, for some combinations of fleet size and production rate, substantial gains will accrue over the short term if the gap is extended.

- Extending ship life to 35 years would probably result in modest savings in discounted terms (perhaps on the order of $2 billion) over the next several decades, regardless of the fleet size or production rate. However, we do not in this estimate account for any costs of determining the feasibility of ship life extension or any costs necessary to effect such extensions beyond those of a standard overhaul.

ON THE RISKS OF AN EXTENDED GAP

- In extending the production gap, DoD would run several risks that could add to the delays and costs we have been able to estimate. The industrial base may lose the expertise of individuals and the capabilities of firms that are essential for efficient reconstitution following a gap. It may be difficult for those design and production workers who do remain to integrate all the technologies that become available in the interim into high-performance submarines. And environmental and nuclear regulatory impediments could add years to the time required to reconstitute.

- There can be little tolerance for trial and error in nuclear-submarine design and construction. Losses of individual and institutional expertise could raise the risk of system malfunction and of an accident, possibly a nuclear one that would entail grave consequences.

RECOMMENDATIONS

- Considering that the savings from extending the current production gap are uncertain and that the risks of doing so are great, we recommend that construction on the next submarines begin as soon as practicable.

- Specifically, we recommend, first, that the third Seawolf-class submarine (SSN-23) be started around 1996 and that the first submarine of the next class be started as soon as feasible, around 1998.

- Finally, considering that savings may be realized by extending the life span of many of the current class of submarines, we recommend that the Navy carefully evaluate this option.

Our recommendations are based on our own judgment regarding prudent weights to be attached to the results of our quantitative cost and schedule analysis and our qualitative risk assessment. Others using the same methodology would arrive at a different course of action if they took either (or both) of two alternative viewpoints. First, in reaching a restart decision, they might regard the risks as much less important. This would be more defensible over the short run (e.g., in deciding not to proceed with SSN-23) than over the long run (e.g., in postponing NSSN restart into the next decade). Second, they might attach much greater weight to the short-term savings of the maximum-gap strategies. The latter approach might be taken by someone who had little or no confidence in cost projections running 20 or 30 years into the future and who thus heavily discounted future costs.

DESIGN CONSIDERATIONS

Although the principal topic of this report is submarine production, we recognize that without design, there would be no production. Like those needed for production, the special skills, tools, and experience required for submarine design face a risk of deterioration during a protracted gap. In Appendix A, we briefly describe the evolution of submarine design over the years, the current phases of design, and the critical skills needed to sustain design capability and what is required to retain them.

For convenience, we often use the term *design* to encompass engineering. To be precise, design is the creative activity encompassing naval architecture and all aspects of marine engineering necessary to produce a new concept or design or a major modification to an old one. Engineering is the application of engineering tools and principles to solve specific problems for the designer and to support the translation of the design to production.

HISTORICAL PERSPECTIVE

The design process has evolved in parallel with the progression of more and more complex submarines, as well as with the changing political and economic environment that has controlled the acquisition of major weapon systems. Prior to the mid-1960s, it would typically take five or six years to design and build the first ship of a class. Subsequently, more extensive reviews for cost and military effectiveness were required and a more thorough definition of the submarine was necessary before issuing a request for proposals. Earlier involvement of the prospective shipbuilders also contributed to a significant increase in the level of effort and time required to complete the design. Finally, of course, the performance requirements for successive classes of submarines have usually been higher, so weapons and other systems have grown more complex. All these factors contributed to progressively increasing submarine class design times. The *Los Angeles* (the first ship of its class) took roughly 12 years to design and build; the *Ohio,* also about 12; and the *Seawolf,* 15 years.

The three-year difference between the last two SSN classes is particularly note-worthy. Specific factors contributing to that increase were as follows:

- The 688 was designed to maximize speed, whereas the Seawolf designers were to maximize performance in all areas.

- Seawolf construction was designed to take advantage of the new modular approach to submarine building, whereas the 688s were originally designed for conventional construction techniques.

- The 688s utilized proven technologies, whereas the Seawolf design had to allow for the use of advanced technologies under development concurrently with the submarine.

DESIGN PHASES

The design process for submarines, like that for other complex warships, consists of four phases.

Concept Design

First, design concepts are explored against the backdrop of a continuing evaluation of future missions, future threats, and future technologies. The dialogues between the designer and the technologist are crucial in this stage. Systems designers with a broad view of technology, current and future operational planning, and operational experience with current designs are necessary to provide leadership. As concepts are explored and defined, tradeoffs are made among military effectiveness, affordability, and producibility. The output of the concept design phase is the definition of a preferred design concept in a set of "single sheet" characteristics that stipulate submarine missions, principal operating and performance characteristics and dimensions, military payload, and design affordability and producibility goals. The cost of construction is also estimated.

Preliminary Design

Next, the preferred concept is matured. Subsystem configurations and alternatives are examined, and structures, hydrodynamics, silencing, and combat system performance and arrangements are analyzed and tested. The output of preliminary design is a set of top-level requirements explicitly describing the refinements achieved during this phase. Performance characteristics and ship dimensions are spelled out in more detail than in the "single sheet" characteristics, and the cost estimate is refined.

Contract Design

The top-level requirements are now transferred into contracts for the detail design and construction of the submarine. All systems must be defined, analysis and testing completed, and the contract drawings for submarine construction prepared. This enables a Request for Proposal (RFP) to be issued so that shipbuilders can respond with proposals that form the basis for the negotiation of the price, terms, and conditions in the final contract.

Detail Design

Normally, the shipbuilder turns the contract drawings and ship specifications into the documents necessary to construct, outfit, and test the ship. Typical products would consist of working drawings, work orders, test memoranda, shipyard procedures, erection sequences, and the like. This phase of design extends into the construction period.

Further Evolution of the Process

For the new attack submarine (NSSN), greater emphasis will be placed on minimizing program cost and risk. Cost estimating must be based on detailed system inputs, requiring a far greater degree of definition earlier in the process. Contract design will concentrate on detailed system definition. Technology must be assessed carefully, with emphasis on minimizing cost and risk.

CRITICAL SKILLS AND PRODUCTION RESTART

There would be obvious advantages in embarking on a refined design process with an experienced team. Even more important are the skills built over the years, skills that must overcome challenges not faced in the design of other ships, such as operating in three dimensions, submerging and surfacing, and integrating systems to meet strict weight and volume limitations. These skills are listed in Table A.1.

Though not listed in the table, some of the most important participants in the design process are those who must lead and oversee the effort. The responsibility for total submarine design synthesis rests with a few individuals—some civilian, some military. Without this talent, the other specialists will diligently work to carefully design and engineer systems and parts that when combined are not likely to function well as a whole or to be easily producible.

If the submarine skill base is dispersed through a period of inactivity, reconstitution will be difficult and time-consuming. It may take several years to train

Table A.1

Critical Engineering and Design Skills

Submarine hydrodynamicists
Submarine hydro-acousticians
Submarine weight engineering
Submarine weapon system engineers and designers
Submarine acoustics engineers and designers
Unique piping system engineers and designers
Submarine naval architects
Submarine propulsion engineers and designers
Submarine electrical system engineers and designers
Submarine combat systems engineers and designers
Submarine structural engineers
Underwater shock engineers
Magnetic silencing engineers
Submarine construction and production engineers
Submarine controls engineers (ship control)

engineers and designers to be effective, and that assumes there are sufficiently skilled personnel available to do the training. The leadership functions would require at least ten years to redevelop.

To avoid such an outcome, it would be necessary to keep 2000–3000 designers and engineers employed in the private sector and at the Naval Sea Systems Command. Ongoing submarine-related research and development (R&D) funding from the Navy and the Department of Energy (DOE) provides support to Navy, contractor, and university technical communities, but lends little sustenance to the activities traditionally associated with new submarine design. If the NSSN design effort does not continue unbroken, the number employed in submarine design will drop below the necessary minimum unless mitigating measures are taken.

One possible mitigating measure is to design and build a prototype. Such an action would ensure that the design and engineering capability would last at least through detail design and would afford designers an opportunity to incorporate new technologies and producibility concepts. Indeed, DoD should seriously examine the possibility of periodic prototyping as a means of "tiding over" design and engineering talent in the gaps between increasingly infrequent new ship classes.

SUPPLEMENTARY BACKGROUND INFORMATION

Appendix B supplements the background information in Chapter Two. We characterize the world nuclear-powered fleet, including the status of national submarine fleets. We review submarine roles and missions past and present and display the course of U.S. submarine construction funding over the years.

SUBMARINES IN THE WORLD NUCLEAR FLEET

Ships powered by nuclear propulsion systems have a relatively long history extending over nearly 40 years from the launching and commissioning of the submarine USS *Nautilus* in 1954. Currently, there are about 350 nuclear-powered ships and submarines. Except for eight icebreakers and a cargo ship belonging to members of the Commonwealth of Independant States (CIS), all are naval vessels. Over half belong to the former Soviet Union, and over a third to the United States. Only the United States and the CIS operate nuclear-powered surface ships, and only the CIS operates civilian manned ones. The United Kingdom, France, and China have the remainder. Submarines make up over 90 percent of the world's nuclear-powered ships (see Figure B.1).

Forty-three nations operate submarines that range from battery-powered midgets through conventional diesel-electric submarines to behemoth nuclear-powered ballistic-missile-carrying submarines. Table B.1 summarizes the submarine fleets throughout the world.

Five countries have built and operated nuclear-powered submarines. Great Britain, France, China, Russia, and the United States all operate both SSNs and SSBNs in their navies.

Russia has been the largest producer of submarines. Today it has the largest and most varied fleet, but has experienced a period of decline as a result of economic difficulties that have impacted operating and maintenance funds. Russia has also exported a large number of submarines to other nations, particularly diesel-electric submarines of the Romeo, Foxtrot, and Kilo classes.

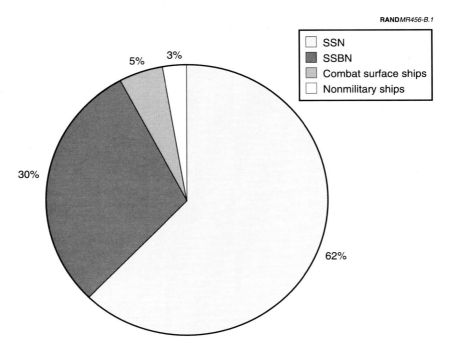

RAND*MR456-B.1*

Legend:
- SSN
- SSBN
- Combat surface ships
- Nonmilitary ships

5%   3%

30%

62%

Figure B.1—Nuclear-Powered Ships by Type

The list in Table B.1 includes long-time allies of the United States as well as potential global and regional threats that are likely to help define the nature and size of the U.S. attack submarine fleet in the decade to come.

ROLES AND MISSIONS

Submarines have played a critical role in military operations for most of this century. Since their introduction as a military weapon during World War I, the submarine's importance in naval warfare has grown substantially. In their initial decades, submarines were found primarily in anti-surface-warfare applications. Their armament, consisting of torpedoes and deck guns, was brought to bear against merchant shipping and enemy warships. The German submarine force during World War II nearly strangled the Allied supply effort across the Atlantic, sinking over a thousand ships. The awesome loss of life and material demoralized the early Allied war effort. Soon thereafter, the American submarine fleet in the Pacific severed vital Japanese resupply routes. By carrying the fight to the enemy's homeland waters early in the war, U.S. submarines starved Japanese industry, paving the way for Allied forces to achieve victory.

Table B.1

Submarine Fleets of the World

Nation	Number of Submarines (by Class)
China	1 SSBN (Xia); 1 SSB (Golf); 5 SSNs, plus 1 under construction (Han); 1 SSG (Romeo); and 43 SSs, plus 60 in reserves and 2 under construction (Romeo, Whiskey, Ming)
France	5 SSBNs, plus 2 under construction and 2 planned (L'Inflexible, Le Triomphant); 6 SSNs, plus 1 in storage (Rubis); and 8 SSs (Agosta, Daphne)
Germany	18 SSs, plus 12 planned (212, 206)
India	18 SSs, plus 1 under construction (Kilo, Foxtrot, Type 209)
Italy	8 SSs, plus 4 under construction (Sauro, Toti, Type S90)
Japan	17 SSs, plus 2 under construction and 1 planned (Yuushio, Uzushio, Harushio)
Libya	12 SSs (Foxtrot, R-2 Mala)
North Korea	24 SSs, plus 2 under construction (Romeo, Whiskey), and 48 midgets
Norway	12 SSs (Ula, Kobben)
Russia	59 SSBNs (Delta, Yankee, Typhoon); 38 SSGNs, plus 3 under construction (Charlie, Echo, Oscar, Yankee); 72 SSNs, plus 7 under construction (Victor, Akula, Alfa, Sierra, Yankee Notch, Yankee, Uniform, Paltus); 12 SSGs (Juliett); and 87 SSs, plus 5 under construction (Foxtrot, Kilo, Tango, Bravo, India, Lima, Beluga, X-ray, Losos)
South Korea	7 SSs, plus 6 under construction and 3 planned (KSS-1, Cosmos, Type 209)
Sweden	12 SSs, plus 3 under construction and 2 planned (Vastergotland, Nacken, Sjoormen, Gotland)
Turkey	15 SSs, plus 2 under construction and 4 planned (Guppy, Type 209, Tango)
United Kingdom	4 SSBNs, plus 3 under construction and 1 planned (Resolution, Vanguard); 13 SSNs (Trafalgar, Swiftsure, Valiant); and 6 SSs, plus 1 under construction (Upholder, Oberon)
United States	27 SSBNs, plus 4 under construction (Ohio, Franklin); 87 SSNs, plus 11 under construction (Los Angeles, Sturgeon, Narwhal, Seawolf); 1 SS (Dolphin); and 1 research sub (NR-1)

SOURCE: *Jane's Fighting Ships 1992–93*, Brassey's Ltd., London.

NOTE: Others include Albania (2 SSs), Algeria (2 SSs, plus 2 planned), Argentina (4 SSs, plus 2 under construction), Australia (5 SSs, plus 3 under construction and 3 planned), Brazil (4 SSs, plus 2 under construction), Canada (3 SSs), Chile (6 SSs), Colombia (2 SSs and 2 midgets), Cuba (3 SSs), Denmark (5 SSs), Ecuador (2 SSs), Egypt (8 SSs), Greece (10 SSs), Indonesia (2 SSs), Iran (2 SSs, plus 1 under construction, and 2 midgets), Israel (3 SSs, plus 2 under construction and 1 planned), Netherlands (5 SSs, plus 2 under construction), Pakistan (9 SSs), Peru (10 SSs), Poland (3 SSs), Portugal (3 SSs), Romania (1 SS), South Africa (3 SSs), Spain (8 SSs), Syria (3 SSs), Taiwan (4 SSs), Venezuela (2 SSs), and Yugoslavia (5 SSs and 6 midgets).

Submarine type abbreviations—"SS" is diesel-electric attack submarine; "B" indicates ship carries ballistic missiles; "G" indicates ship carries guided missiles; and "N" indicates nuclear power.

The traditional missions of submarines expanded greatly after World War II. Most obviously, submarines came to provide the nation with the credible, survivable, mobile nuclear deterrent that formed the backbone of U.S. security strategy. Their sonar and combat system capabilities were so greatly improved that they became the premier *anti*submarine force in the U.S. military and the only forces that could conduct operations under the North Atlantic and Arctic ice cap.

These traditional roles have become less critical with the end of the cold war. Other missions, however, demonstrate the utility of the attack submarine in a new security environment characterized by a multiplicity of unpredictable, seemingly minor situations that might warrant bringing military force to bear. The nuclear propulsion technology with which all U.S attack submarines are now equipped is particularly suited to the demands of the era because of the high submerged speed and extended endurance it permits. Submarines can be rapidly sortied to respond to a crisis and remain in a ready condition for months without the need to resupply, refuel, or return to base. The submarine's stealth allows it to remain in a crisis area undetected, maintaining surveillance over the situation as necessary. If the crisis is defused by nonmilitary means, the submarine can be withdrawn without political embarrassment or other implications—perhaps for rapid redeployment to another crisis situation. If action is needed, the submarine is on station ready to perform a variety of missions:

- Deep strike. Submarines carry highly capable cruise missiles whose long range holds 75 percent of the earth's surface at risk. Two nuclear submarines successfully struck key Iraqi targets with Tomahawk missiles during the Persian Gulf War.

- Anti-surface-ship warfare. Modern torpedoes, antiship missiles, and mines endow today's submarines with a formidable capability in this traditional submarine role.

- Mine detection. Improved sonar systems provide the attack submarine with the capability of detecting and mapping potentially mined areas. This information could be critical to other naval forces and for the resupply of ground and air forces.

- Insertion of special forces. All U.S. attack submarines have this capability, as do two SSBNs converted for this purpose. Submarine stealth and shallow-water operating capability facilitate this mission.

U.S. SUBMARINE CONSTRUCTION FUNDING PROFILE

A historical profile of U.S. submarine production is offered in Chapter Two. Another perspective can be gained by examining the course of total funding for submarine construction, as is authorized in the budget account Shipbuilding and Conversion, Navy (SCN). This account also includes construction funding for all other Navy vessel construction. Figure B.2 shows the early-1960s peak in submarine construction that is evident in Figure 2.6. It shows a secondary peak in FY73 that is not evident in Figure 2.6 because construction starts for submarines authorized in the early to mid-1970s were spread over several years thereafter. Finally, the graph shows the drop-off in submarine construction funds to a zero obligation in FY93.

Submarine construction funding has also dropped as a percentage of total ship construction. In most years through the mid-1970s, submarine construction funding was at least a third of that for all ship construction, in some years half. Over the last 15 years, submarine construction has usually been under a quarter of the SCN budget.

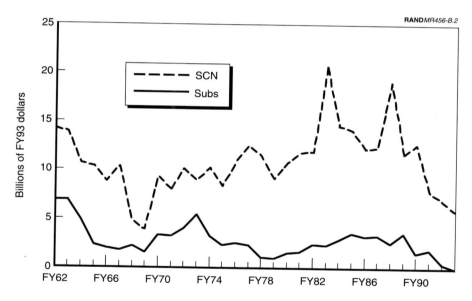

Figure B.2—Submarine Construction Funding vs. Navy Ship
Construction Funding (SCN)

SHIPYARD EFFECTS: ADDITIONAL CASES

In Chapter Three, summary estimates were given for cost and schedule effects of gaps in submarine production on the shipyards. For illustration, the underlying details were shown of the cost estimate for one of the cases examined—Electric Boat (EB) without further work after currently programmed submarine production ends. Here, we present the cost estimate details for the other five cases: Newport News Shipbuilding (NNS) without further work beyond current submarine and carrier construction; NNS, with CVN-76; Electric Boat, with submarine overhaul work; NNS, without CVN-76 but with submarine overhaul work; and Electric Boat with SSN-23. We conclude with a tabular summary of all cases.

BASELINE ESTIMATES FOR NEWPORT NEWS SHIPBUILDING

In this section, we give our cost and schedule estimates for NNS without the funding of CVN-76 in the foreseeable future.[1] These estimates are in some respects more complex than those for EB because one shop may be working not only on submarines but also on other ships and because personnel may move from one line to another. We simplify matters somewhat by restricting ourselves to submarine and carrier work only. This may lead to some misestimates—for example, if workers no longer needed in submarine construction move within the yard to help in building or overhauling other military or commercial ships. In that case, our estimates of the costs to release, rehire, and retrain personnel may be overestimated, as skilled submarine production workers may be retained in the yard and be available when submarine production resumes.

---

[1]The Navy plans a carrier overhaul in FY03–04. We ignore the manpower requirements for that overhaul for two reasons. We do not analyze submarine restarts beyond FY04 (when these overhaul workers would become available for submarine construction) in this appendix, and, as we demonstrate in Chapter Six, restarts after FY01 are impractical.

Impact on Submarines Currently in Construction

At the end of FY93, NNS had about twice as many people employed in carrier work as in submarine construction. Although both workforces are declining, the types of needed skills change over the course of carrier construction, and released submarine workers may find work on the carrier line. We draw the optimistic conclusions that, with prudent yard management, few workers will have to be laid off from the yard before the current submarine production run ends in FY96 and that morale-related productivity losses can be minimized. We therefore assume there will be no impact on the costs of the submarines currently in construction if further submarine production is delayed, even for several years.

Costs of Smart Shutdown

NNS estimates that, because of other work in progress, only a quarter of the submarine production facilities would go completely out of service should the submarine line shut down; others would be partially shut down. Mothballing, however, might not be necessary in all cases, as some activities formerly occurring in shut-down facilities might be resumed at restart in facilities remaining open.

However, for restarts after carrier work ends (FY98 or later), smart shutdown will be necessary at all facilities. For these, we assume that NNS facility and equipment shutdown costs would be comparable to EB's. For submarine restarts in FY95 through FY97, we assume such costs would amount to 20 percent of EB's costs (for later restart years; NNS's submarine work ends sooner).

With regard to nuclear operations and fuelings and vendor liability, ongoing carrier work should ensure against any costs resulting from the end of submarine production at least through the mid-90s. For restarts after that, we assume NNS's costs are comparable to EB's.

The last element of cost associated with stopping submarine production at NNS is related to personnel. As was the case for impact on current construction, we assumed we could neglect these costs. Total facility, equipment, and vendor costs for shutdown at NNS are given in Table C.1.

Annual Cost of Maintaining NNS Production Capabilities

As we have no detail from NNS on the cost to maintain their capabilities during a production gap, we adapt EB's later-year values (see Table C.2). Carrier work holds some maintenance costs to 20 percent of their maximum value for restarts before FY98.

Table C.1

**Cost to Shut Down NNS Submarine
Production Capability
(millions of FY92 dollars)**

Restart Year	Total
FY95	8
FY96	8
FY97	25
FY98 or later	68

Table C.2

**NNS Annual Maintenance
Costs Prior to Restart
(millions of FY92 dollars)**

Year	Added cost
FY95	5
FY96	5
FY97	5
FY98 or later	46

NOTE: Costs on a given line accrue if restart occurs in that year or afterward.

Costs and Schedule of NNS Reconstitution

For NNS's equipment and facility reconstitution costs, we use EB's values, with
the 20 percent factor applied for restarts before FY98. We calculate the cost of
reconstituting NNS's workforce much as we did for EB's. Data supporting
workforce inputs to the workforce reconstitution model are shown in Table C.3.
Note here, however, that "workers remaining" include those on both the submarine and carrier lines. As for EB, we assume that for restart in a given year,
NNS can rehire 90 percent of the workers released the previous year (this time
from both the submarine and carrier lines) and 20 percent of those released the
year before that. The total is the initial workforce at restart, given as "workers
available" in Table C.3. (Note we constrain this again to a minimum of 260
workers.) We also assume that the reconstituting workforce absorbs 90 percent
of personnel coming off other construction lines in the year of restart and in
subsequent years (these are the "skilled transfers" in the table).

We add two elements of cost to those generated from the model. For the cost to
hire support personnel, we make the same adjustment we did for EB. But here,
we also allow $1600 to provide remedial training to each worker moving from
CVN to SSN production. We assume that these persons have worked on the

Table C.3

Skilled Workforce Available When Production Resumes

Restart Year	Workers Remaining at Start of Year[a]	Workers Released During Previous Year[a]	Workers Available for Restart	Skilled Transfers This Year and After
FY95	10000	3000	2700	9000
FY96	6300	3700	3930	5670
FY97	3300	3000	3440	2970
FY98	1300	2000	2400	1170
FY99	260	1040	1336	0
FY00	260	0	468	0
FY01 or later	260	0	260	0

[a]Including those on CVN line.

submarine line at some time but may not have retained submarine skills. The $1600 figure is drawn from an analysis based on an NNS study[2] of the company's ability to reconstitute the submarine construction work force.

For that study, NNS identified the basic work skills, formal worker qualifications, and complex work operations required in each submarine shop. NNS also identified which of those would be required in carrier construction. The study results suggest 92 percent of the basic submarine skills, 82 percent of the qualifications, and 69 percent of the complex operations would be maintained through carrier construction.

For each of the shops, we took the difference between the number of skills, qualifications, and complex operations required for submarine construction and the number maintained by carrier construction. We multiplied those differences by the peak quarterly number of personnel required for submarine construction to get the total number of training "actions" required for each shop and summed these across all shops. We estimated $500 per training action and divided the total cost by the number of workers needed to build a submarine to get $1600 to train each carrier transfer. To give a graphic impression of the similarities and dissimilarities in the skills required, Figure C.1 compares, for some 30 skills, the number of persons required in the maximum quarter of submarine and carrier construction.

---

[2]*Reconstitution of the Submarine Production Workforce*, Newport News Shipbuilding, April 1993.

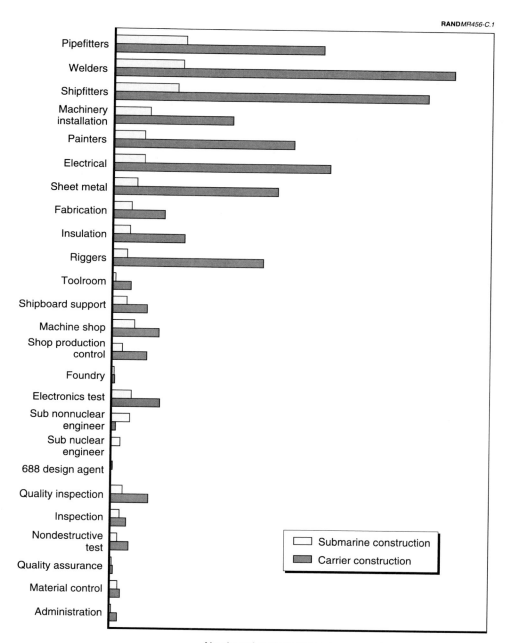

Number of workers in maximum quarter

Figure C.1—Persons Required, Submarine and Carrier Construction, Maximum Quarter

The total of all adjustments ranged from $7 to $13 million, depending on the restart year and production rate. By adding these to the model outputs, we obtain our estimate of the total cost to reconstitute the production workforce. These estimates for various production rates are shown in Table C.4 and Figure C.2, for the assumptions given in the Note to the table. Again, costs are high and increase with gap length, from several hundred million dollars in FY98 to the neighborhood of $2 billion if restart is delayed beyond FY00.

Gap length affects the postrestart production schedule in much the same way it does for EB (see Figure C.3, assumptions as for Table C.4). Deferral of restart beyond FY99 increases the time to deliver the first ship, resulting in a delay of several years for restarts later than FY00. And the delay is at least twice as great for achieving a sustained rate of production.

Summary Estimates

Figure C.4 shows the various costs associated with submarine production gaps at NNS if CVN-76 is not funded and the eventual sustained production rate is two ships per year. Total costs grow rapidly with gap length, driven by personnel-related costs.

NEWPORT NEWS WITH CVN-76

If CVN-76 is funded, Newport News will be able to continue to use a portion of the submarine personnel and facilities beyond the time when the current carrier work ends in 1998. Here, we briefly summarize the costs of restart with CVN-76. Most of these are small and do not vary with restart year.

Table C.4

Total NNS Personnel Reconstitution Costs Without CVN-76 (millions of FY92 dollars)

Restart Year	Rate = 2	Rate = 3
FY98	416	705
FY99	750	1099
FY00	1453	1799
FY01 or later	1747	2125

NOTE: Case shown assumes fixed overhead of $150 million (pessimistic), early attrition rate of 5 to 10 percent (optimistic), and mentor: trainee ratio of 0.5 (intermediate).

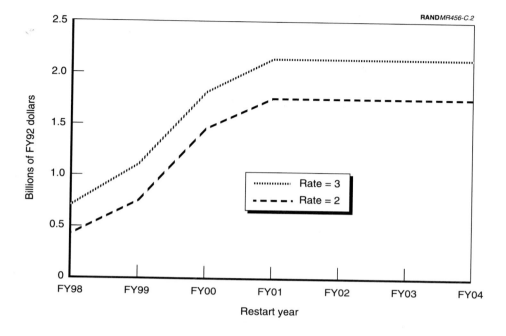

Figure C.2—NNS Personnel-Related Reconstitution Costs, Without CVN-76

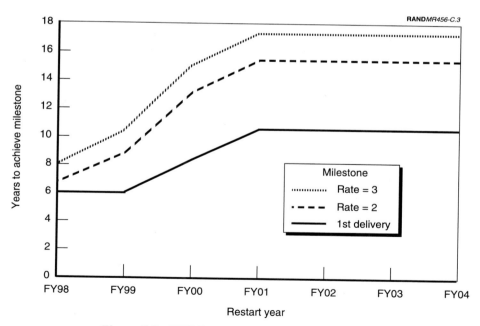

Figure C.3—NNS Postrestart Production Schedule, Without CVN-76

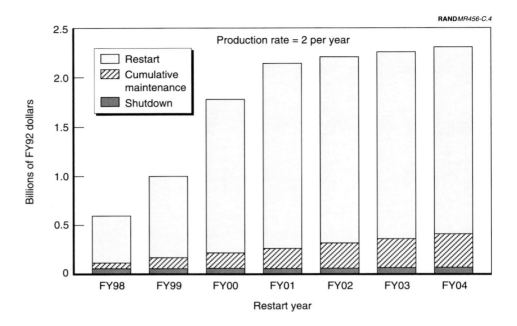

Figure C.4—NNS Shipyard Reconstitution Costs, No Work Beyond That Currently Under Way, Production Rate = Two per Year

Since we reasoned there would be no effect on the current production submarines if CVN-76 is not funded, we infer there would be none if it is. Shutdown costs would be equal to those in the previous case for restart prior to carrier line shutdown—$8 million. The additional carrier work will keep the nuclear operations active, and we expect no vendor liability costs or costs for releasing personnel.

Annual maintenance costs are also equal to those in the previous case for restart prior to carrier line shutdown. These amount to $5 million—the annual cost of security and maintenance manpower and materials, utilities, emergency repairs, taxes, and insurance. There is also no need for a cadre to ensure the availability of mentors when submarine production is restarted.

Since few of the submarine facilities would be completely mothballed, there would be only minor costs associated with bringing them back on-line. We estimate $5 million to reconstitute facilities and $2 million for equipment.

To calculate workforce reconstitution costs, we again establish an initial workforce from released submarine and carrier workers. In contrast to the previous case, however, the carrier workforce is growing during some potential restart years (FY98 through FY00; see Figure C.5), so fewer workers are available for restarting the submarine line. For cases when those workers number less than 1000, NNS indicated they would move skilled submarine workers from the CVN line to the submarine line and hire new replacement workers on the CVN line.[3] The resulting initial workforce numbers we input to the workforce reconstitution model are shown in Table C.5.

We make the same adjustments described in the previous case for support personnel and retraining CVN workers, and we cost replacements for CVN workers moved to the submarine line at $10,000 each. (We could not quantify losses to the carrier line arising from the inefficiency of the replacement workers, but they may be in the thousands of dollars per worker.) The total of all adjustments ranges from $5 to $26 million.

The total cost of reconstituting the workforce is shown in Table C.6 and Figure C.6. Note that, following the jump from FY98 to FY99, personnel-related reconstitution costs decrease with a longer gap. The reason is that a large number of transfers from CVN-76 is available in FY04. Thus, the closer the restart

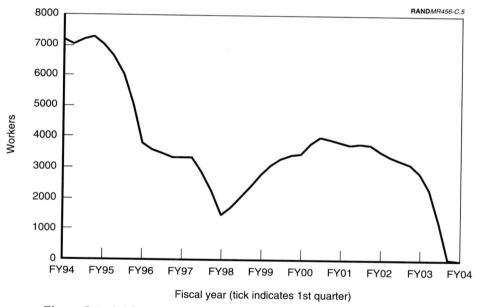

Figure C.5—With CVN-76, Carrier Work Increases in Some Submarine Restart Years

---

[3]Note that we do not let the workforce fall as far in this case as we did in either baseline case. The 260-person cadre was kept small because of its cost and the difficulty of finding tasks that would maintain the skills of many more people. But in the present case, a large number of workers whose skills are maintained by productive work is available. We believe it likely that the shipyard would take advantage of that to restart submarine production with a larger workforce and reap major cost advantages.

Table C.5

Skilled NNS Workforce Available When Production Resumes If CVN-76 Is Funded

Restart Year	Workers Remaining at Start of Year[a]	Workers Released During Previous Year[a]	Workers Available for Restart	Workers Hired from CVN[b]	Skilled Transfers This Year and After[c]
FY95	10000	3000	2700	0	10570
FY96	6300	3700	3930	0	7200
FY97	3400	2900	3350	0	4590
FY98	2200	1200	1660	0	3510
FY99	2400	0	40	960	3510
FY00	3400	0	0	1000	3510
FY01	3900	0	0	1000	3510
FY02	3800	100	90	910	3420
FY03	3100	700	650	350	2790
FY04	260	2840	2956	0	0

[a]Including those on CVN line.

[b]To bring initial submarine restart workforce to 1000.

[c]Exceeds workers remaining in some restart years because additional workers are hired for CVN in subsequent years.

Table C.6

Total NNS Personnel Reconstitution Costs with Funding of CVN-76 (millions of FY92 dollars)

Next Start	Rate = 2	Rate = 3
FY98	479	870
FY99	746	1019
FY00	675	932
FY01	588	845
FY02	510	764
FY03	453	735
FY04	446	756

NOTE: Case shown assumes fixed overhead of $150 million (pessimistic), early attrition rate of 5 to 10 percent (optimistic), and mentor:trainee ratio of 0.5 (intermediate).

year is to FY04, the shorter the time over which the force must be "grown" from scratch.

The relationship between restart year and schedule also differs from that of previous cases (see Figure C.7). The larger initial submarine workforces permit-

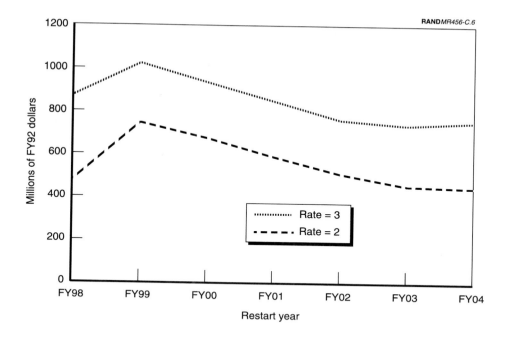

Figure C.6—Personnel-Related Reconstitution Costs for NNS with CVN-76

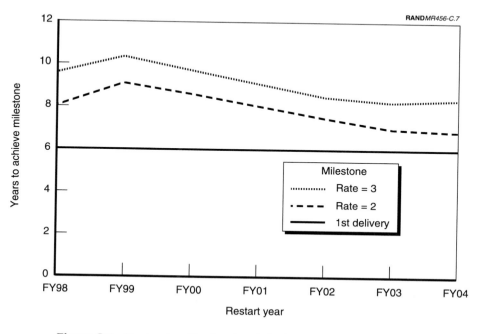

Figure C.7—Postrestart Production Schedule for NNS with CVN-76

ted by transfers from the carrier workforce allow first delivery within six years, regardless of restart year. A sustained production rate of two per year can be achieved within nine years with restart in FY99; after that, the interval diminishes, for the same reason that cost diminishes.

Summary Estimates

Total costs of submarine production gaps at Newport News with CVN-76 are shown in Figure C.8. Those totals are smaller than—and, for restart years beyond FY99, only a fraction of—the totals for Newport News without CVN-76 (plotted as a line in the figure). Thus, taking workers from the carrier line may cause losses there in the millions or low tens of millions of dollars, but the gain on the submarine line is in the hundreds of millions.

IMPACT OF DIRECTING OVERHAUL WORK TO PRIVATE YARDS

We have seen a large difference in the costs associated with an active and an inactive yard because of the variation in the number of people that must be recruited, hired, and trained when submarine production restarts. There appears to be a substantial advantage to keeping a shipyard and its employees busy with

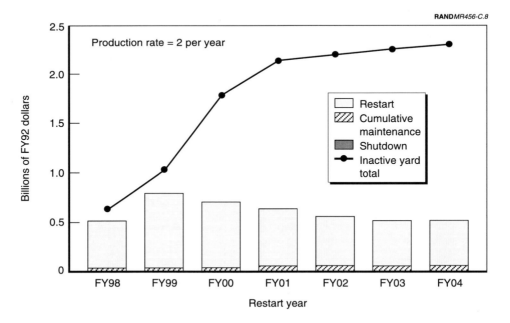

Figure C.8—Shipyard Reconstitution Costs, Newport News, with Additional Aircraft Carrier (CVN-76), Production Rate = Two per Year

work related to submarine construction. A potential way to do so is to direct submarine repair, overhaul, and refueling workload to private shipyards.

The maintenance actions for submarines were described in Chapter Two. The annual repair and overhaul workload for submarines can be substantial. As shown in Table C.7, submarine maintenance actions have averaged about 2.4 million worker-days annually over the last seven years.

It is true that the maintenance workload has been sharply declining over the last several years as the pre-688-class submarines reach the end of their useful life. But it will rise again as the current fleet reaches its programmed force strength and the boats reach their scheduled maintenance points (see Table C.8).

Table C.7

Major Maintenance Activities for Submarines

Year	Pre-688 Class Number	Man-days (000)	688-Class ROHs Number	Man-days (000)	688-Class DMPs Number	Man-days (000)
FY86	10	3669	3	1141	0	0
FY87	10	3578	1	481	0	0
FY88	6	2297	1	450	0	0
FY89	5	1622	1	462	5	833
FY90	2	713	0	0	4	582
FY91	0	0	0	0	4	641
FY92	0	0	0	0	2	283

SOURCE: Naval Sea Systems Command, Management Group, SEA 072.
NOTE: ROH-regular overhaul; DMP-depot modernization period.

Table C.8

Scheduled Future Repair/Overhaul Activities for 688-Class and Ohio-Class Submarines

Year	688-Class DMPs	RFOHs	Ohio-Class ROHs
FY95	0	1	0
FY96	4	1	0
FY97	1	3	2
FY98	3	3	1
FY99	2	3	1

NOTE: DMP-depot modernization period; RFOH-refueling overhaul; ROH-regular overhaul. Schedule assumes no early decommissioning.

The maintenance actions shown in Table C.8 each require between 150,000 and 500,000 worker-days of effort, or approximately 600 to 2000 worker-equivalents. Thus, there should be enough maintenance work for 7000 to 9000 shipyard workers—almost equivalent to the workload for steady-state construction of two new submarines per year. Even if the Los Angeles-class is decommissioned instead of refueled at midlife, there would be enough repair and overhaul work for 2000 to 3000 workers.

There will also be approximately 16 drydocking selected restricted availabilities (DSRAs) per year over the next five years. These periodic maintenance actions, completed between public and private shipyards, will equate to approximately 400,000 worker-days (or 1600 equivalent workers) per year.

There is obviously a potentially significant annual workload in the repair, over-haul, refueling, and decommissioning of submarines. But will this type of work be sufficient to maintain the management and production skills needed for submarine construction?

Management Differences Between Overhaul and New Construction

There are some significant differences between the management techniques and philosophies for new submarine construction and those for repair and overhaul. Construction contracts are Fixed Price Incentive (FPI). Theoretically, the work is well defined in advance, particularly for follow-on ships of a class. It follows a set sequence, rather like the work on an automobile assembly line; un-seen problems are unlikely. Management efforts are directed at ensuring those schedules are achieved as planned.

Overhauls are characterized by large uncertainties in defining the workload ahead of time in terms of both total workload and cost. The work needed is of-ten not known with specificity until systems are disassembled and inspected. Contracts are therefore written as Cost Plus Incentive (CPI).

Because of the uncertainty and because there is pressure to return a ship to the fleet as soon as possible, overhauls are much more management intensive than new construction. Management philosophies and, most important, manage-ment control systems must be more flexible and adaptable. In particular, there are significant differences in management systems for the ordering and control of material, for scheduling workload and assigning skilled workers, for tracking the status of systems and equipment, for processing reports, and for predicting labor requirements.

Although most major overhauls now take place in public shipyards, both EB and NNS have performed overhauls in the past. In those cases, increases in overhaul workload at the two yards were coincident with increases in new construction workload, and both cost and schedule performance degraded significantly.

In a low workload situation, as would result from gaps in new construction, intensive management should be able to accomplish submarine overhauls in a timely and cost effective fashion. However, a period of adjustment, along with some start-up costs, would be required.

Trade Employment in New Construction Versus Overhaul

A potentially more important issue is whether overhaul work involves construction skills. Figure C.9 compares the maximum quarterly worker-hours for submarine construction and overhaul. All but the "pure" production skills are employed in submarine overhauls, but for many skills, fewer hours are required. Shipfitters, welders, and grinders, for example, are used to a much greater extent in new construction. However, outside machinists, electronics installation and test, and engineering and finance are used more in overhaul.

Thus, trade people can remain employed in overhauls, but can their production skills be maintained? Newport News determined that 96 percent of the basic work skills, 99 percent of the qualifications, and 89 percent of the complex work operations needed for submarine construction could be maintained through submarine overhaul.[4] Certainly, training will be required when moving back to construction, but the level of training should be significantly less than that required for a new hire. And keeping submarine personnel employed in overhaul will eliminate the costs of releasing and rehiring those workers.

At NNS, carrier overhaul could also be considered as a way to keep submarine construction personnel employed during the hiatus in submarine production. Figure C.10 compares skill requirements for submarine construction with those for carrier overhaul.

Cost Impact of Directing Overhaul Work to EB

Placing major overhaul and repair work in private yards can have a significant impact on the costs associated with shutting down, maintaining, and reconstituting submarine production capability. Many of the production support facili-

---

[4]*Reconstitution of the Submarine Production Workforce*, Newport News Shipbuilding, April 1993. We estimate that only a quarter as many training actions are needed to move the workforce from submarine overhaul to construction as to move it from carrier construction to submarine construction.

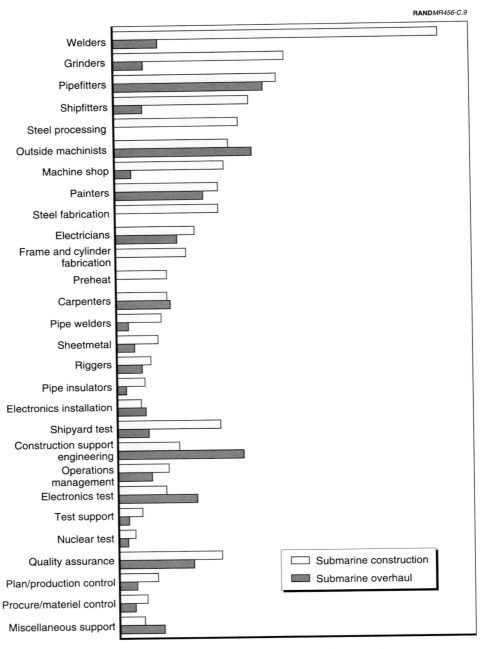

Figure C.9—Submarine Construction and Overhaul Skill Requirements

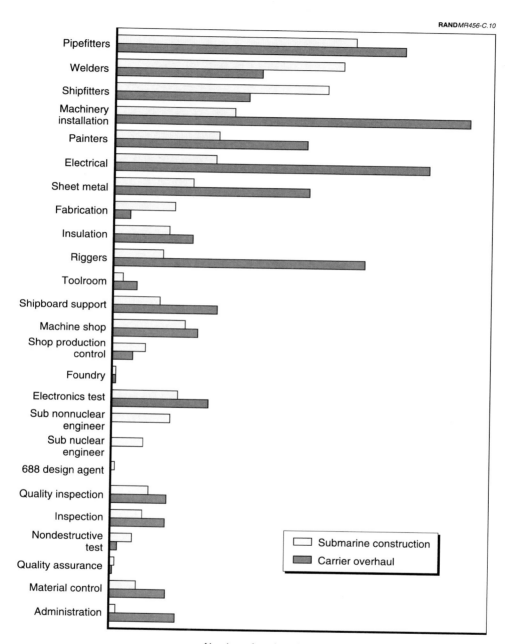

RANDMR456-C.10

Number of workers in maximum quarter

Figure C.10—Submarine Construction and Carrier Overhaul Skill Requirements

ties could be used for overhaul work, and personnel-related costs would be lower.

In estimating that impact for EB, we assume enough overhaul work is assigned to employ 1000 people. This equates to at most two 688-class DMPs per year or one Ohio-class ROH per year. We further assume that the work does not involve refuelings or the handling of "dirty" fuel, which EB cannot now do.

Impact on Submarines in Construction. Overhaul work coming into the yard would result in the release of fewer employees and a decrease in the number of incentive bonuses to keep key people in place until the final boat is delivered. Inefficiencies associated with an approaching shutdown should also be less pronounced. If enough workload is assigned to employ 1000 workers, we estimate costs could decrease by half for the longest gaps considered (compare Table C.9 with Table 3.1).

Shutdown. Overhaul work would keep many of the production support facilities open and equipment in use at the Groton location. Approximately 50 percent of the Groton shutdown costs could be saved (compare Table C.10 with Table 3.3). The nuclear operations/fueling cost element may remain the same, but about 1000 fewer people (the 1000 overhaul workers plus support minus the

Table C.9

Cost Impact on EB Submarines Currently in Construction
with Overhaul Work

Next Start	Number of Bonuses	Extra Worker-years	Total Cost ($M FY92)
FY95 to FY98	0	0	0
FY99	500	300	20
FY00 or later	1000	600	40

NOTE: Each bonus costs $10,000; each extra worker-year costs $50,000.

Table C.10

EB Shutdown Costs, with
Overhaul Work
(millions of FY92 dollars)

Next Start	Total
FY95	3
FY96	18
FY97	58
FY98	78
FY99	84
FY00 or later	115

cadre) would be released, so severance pay and retraining costs would be lowered. Vendor liabilities remain the same as for the baseline case, as submarine production is shut down here also.

Annual Maintenance. Because fewer facilities are shut down, the costs to EB for maintaining the shut-down facilities—costs for security, utilities, taxes, and insurance—are lower. Potentially 50 percent of such costs at Groton could be "saved" (compare Table C.11 with Table 3.4). Also, there is no need to establish a cadre, so there is no personnel cost. Maximum annual savings may approach $32 million.

Reconstitution. Fewer facilities and less equipment mothballed imply less cost in reconstituting the nonpersonnel-related production capability. We assume that the break-out cost for the Groton facilities and equipment and for procedure qualifications could be reduced by 50 percent. We also assume that 50 percent of the computer start-up cost could be saved.

Placing overhaul work in the inactive yard has its biggest impact on the cost to reconstitute the production workforce. More workers stay employed, so fewer new workers must be hired and trained for the start of production. Also, there is a larger base of skilled workers to build upon, so reconstitution should be smoother and quicker.

To estimate the total skilled workforce available when submarine production restarts, we use the workforce drawdown profile from the second column of Table 3.5 (beginning with FY95), but do not let it fall below 1000. We thus assume that overhaul work is directed to the yard in the quantities necessary to support a workforce of 1000. The other columns are adjusted accordingly (see Table C.12). These numbers then are input to the workforce reconstitution model, whose outputs are corrected for hiring of support personnel, a correc-

Table C.11

**EB Annual Maintenance Costs Prior to Restart,
with Overhaul Work
(millions of FY92 dollars)**

Year	Quonset	Groton	Total
FY95 to FY96	0	0	0
FY97	8	0	8
FY98	8	0	8
FY99	8	0	8
FY00 or later	8	8	16

NOTE: Costs on a given line accrue if restart occurs in that year or afterward.

Table C.12

Skilled Workforce Available When Production Resumes

Restart Year	Workers Remaining at Start of Year	Workers Released During Previous Year	Workers Available for Restart	Skilled Transfers This Year and After
FY95	10500	500	550	9450
FY96	8000	2500	2350	7200
FY97	4000	4000	4100	3600
FY98	1500	2500	3050	1350
FY99	1000	500	1450	450
FY00	1000	0	1100	0
FY01 or later	1000	0	1000	0

tion that ranges from $4 to $10 million.[5] Results are shown in Table C.13 and Figures C.11 and C.12.

The results exhibit the expected improvements in cost and schedule, relative to those in an inactive yard, because of the larger initial workforce (compare with Table 3.6 and Figures 3.1 and 3.2). Note, however, that when the 1000 workers kept busy with overhaul work are transferred to new production at restart, *there will be no one left to work on overhauls, and that workload will have to be shifted back to a public yard.* That involves costs either in rehiring workers at those yards or in maintaining a skilled, ready cadre there—costs that we do not take into account.

Summary Estimates. Figure C.13 shows the various elements of costs associated with submarine production gaps at EB when enough overhaul work to

Table C.13

Total EB Personnel Reconstitution Costs,
with Overhaul Work
(millions of FY92 dollars)

Next Start	Rate = 2	Rate = 3
FY98	331	606
FY99	687	1017
FY00	942	1328
FY01 or later	1004	1386

NOTE: Case shown assumes fixed overhead of $150 million (pessimistic), early attrition rate of 5 to 10 percent (optimistic), and mentor: trainee ratio of 0.5 (intermediate).

---

[5]Some retraining will be required for overhaul workers, but it will be much less than that for carrier workers in the previous Newport News case, so we ignore it here.

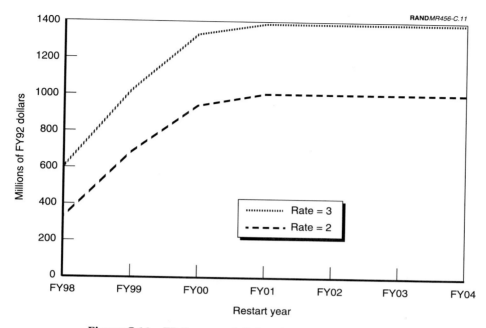

**Figure C.11—EB Personnel-Related Reconstitution Costs,
with Overhaul Work**

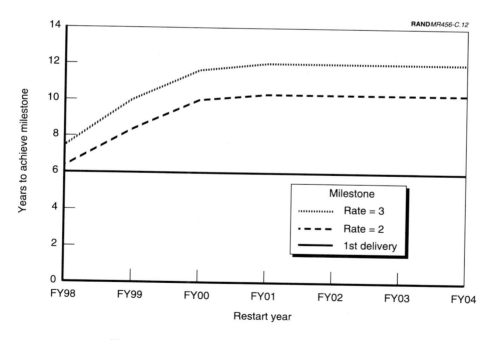

**Figure C.12—EB Postrestart Production Schedule,
with Overhaul Work**

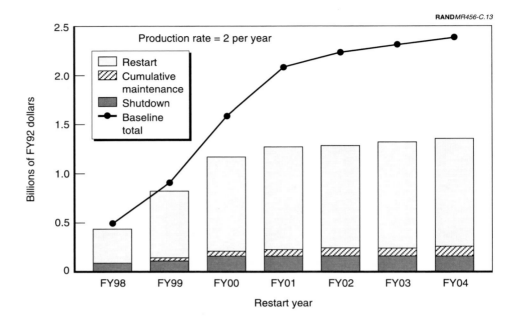

Figure C.13—EB Shipyard Reconstitution Costs, with Overhaul Work, Production Rate = Two per Year

keep 1000 persons per year busy is placed in the yard. Savings relative to gapped production with an inactive yard are on the order of one-third to one-half (or roughly half a billion to a billion dollars) for restarts after FY99.

Cost Impact of Directing Overhaul Work to NNS

We next consider the NNS case in which CVN-76 is not funded but overhaul work sufficient to keep 1000 direct workers employed per year is directed to the yard. Our cost estimates are based on submarine overhauls, although we believe the costs would be similar (perhaps slightly higher to account for retraining) if the work was associated with the overhaul of carriers.

Impact on Submarines in Construction. As with our other cases for NNS, we assume there is no impact on the cost of submarines currently under construction. Prudent yard management should be able to move workers across the submarine and carrier construction and overhaul lines in a manner that would reduce the release of workers and minimize morale-related productivity losses.

Shutdown. Currently programmed carrier work will keep the production support facilities operating through FY97. Submarine or additional carrier overhaul work would keep a portion of the production support facilities open beyond that time. For submarine starts in FY98 or later, we assume the costs at NNS would be comparable to the costs at EB (with overhaul work).

We assume the other elements of shutdown cost (nuclear operations, vendor liabilities, and personnel release) will be the same as in the NNS base case. Table C.14 summarizes our estimates of shutdown costs at NNS when overhaul work is available. NNS can refuel submarines or carriers and, if the overhaul work involves refueling, shutdown costs for starts in FY98 or later could therefore be reduced by the cost associated with shutting down the nuclear fueling operations.

Annual Maintenance. Submarine or carrier overhauls should have no impact on annual maintenance costs for submarine starts prior to FY98 (ongoing carrier work would keep those costs low). For submarine starts in FY98 or later, we assume NNS annual maintenance costs would be comparable to those for EB with overhaul. Our cost estimates for annual maintenance (terminating at restart) are, therefore, $5 million per year through FY97 and $16 million per year thereafter.

Reconstitution. The assumptions that underlie our estimates of the nonpersonnel reconstitution costs mirror those listed above plus those that underlie our EB-with-overhaul case. That is, ongoing carrier work holds down costs associated with submarine starts prior to FY98; submarine overhauls would not further reduce these costs. For submarine starts beginning in FY98, we assume the nonpersonnel reconstitution costs when overhauls are assigned to the yard are 50 percent of the costs associated with no overhauls.

As with EB, the workforce profiles for NNS are the same as the baseline case, but the workforce is not allowed to fall below 1000 (see Table C.15). The workforce reconstitution model's outputs are corrected for support hiring and retraining of carrier workers, which, taken together, add between $8 and $13 million to the model's output. The totals are shown in Table C.16 and graphed in Figure C.14. Schedule effects are shown in Figure C.15.

Table C.14

**NNS Shutdown Costs, with
Overhaul Work
(millions of FY92 dollars)**

Next Start	Total
FY95	8
FY96	8
FY97	25
FY98 or later	53

Table C.15

Skilled Workforce Available When Production Resumes

Restart Year	Workers Remaining at Start of Year	Workers Released During Previous Year	Workers Available for Restart	Skilled Transfers This Year and After
FY95	10000	3000	2700	9000
FY96	6300	3700	3930	5670
FY97	3300	3000	3440	2970
FY98	1300	2000	2400	1170
FY99	1000	300	1670	0
FY00	1000	0	1060	0
FY01 or later	1000	0	1000	0

Table C.16

Total NNS Personnel Reconstitution Costs,
with Overhauls but Without CVN-76
(millions of FY92 dollars)

Next Start	Rate = 2	Rate = 3
FY98	416	705
FY99	728	1068
FY00	968	1357
FY01 or later	1006	1398

NOTE: Case shown assumes fixed overhead of $150 million (pessimistic), early attrition rate of 5 to 10 percent (optimistic), and mentor: trainee ratio of 0.5 (intermediate).

Having overhaul work alone at Newport News does not result in the kinds of improvements in cost or time to achieve sustained production rates (relative to those from gapping with an inactive yard) that can be achieved with CVN-76. The latter serves as a source of thousands of transfers to bolster the reconstituting submarine construction workforce, and we assume only enough overhaul work for 1000 persons. However, those 1000 are enough to hold the time for first ship delivery to the nominal six years, and the improvements in cost and years to achieve sustained production rates are substantial compared to those for an inactive yard.

Summary Estimates. Gap-related costs other than those for reconstituting the workforce are small for all restart years, so the cost conclusions just stated also apply to all gap-related costs (see Figure C.16).

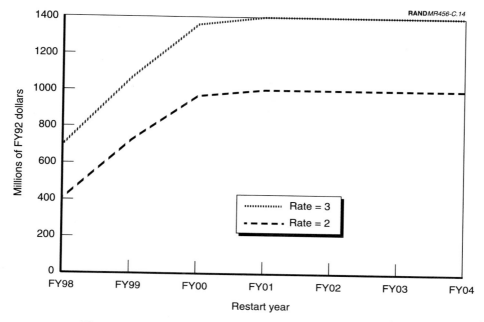

**Figure C.14—NNS Personnel-Related Reconstitution Costs,
with Overheads but Without CVN-76**

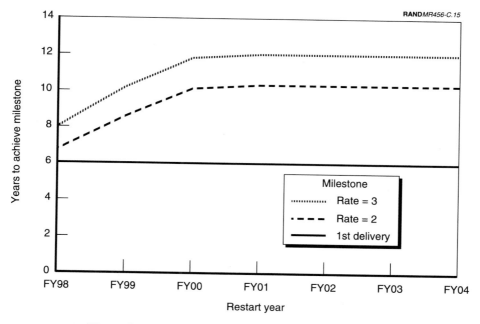

**Figure C.15—NNS Postrestart Reconstitution Schedule,
with Overhauls but Without CVN-76**

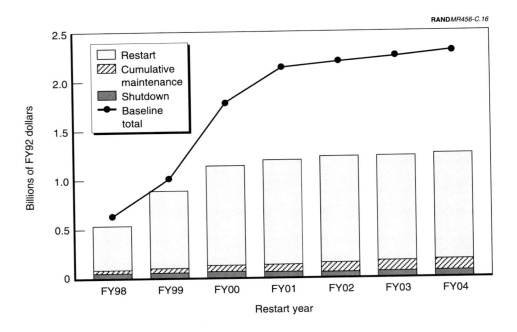

Figure C.16—NNS Shipyard Reconstitution Costs,
with Overhauls but No CVN-76

BRIDGING THE PRODUCTION GAP WITH SSN-23

Placing overhaul work in the private shipyards has the advantage of keeping the shipyard active during the halt in submarine production thereby reducing the total costs. Fewer facilities and equipment must be mothballed, maintained, and reconstituted and, more importantly, workers stay employed during the production gap and provide a larger base from which to rebuild the production workforce. An alternative method for keeping the shipyard active is to build the third Seawolf, SSN-23. Here we consider the case where SSN-23 is authorized for a production start at EB in 1996.

This "bridging" strategy has the following effects on our cost estimates:

- We assume that SSN-23 would provide sufficient workload to eliminate the loss of morale and productivity inefficiencies associated with the submarines currently in construction. That is, we assume that there is no additional cost for the submarines currently under construction.

- The Quonset Point facility stays active until at least 1999 and the Groton facility would not close until 2002. We therefore assume that the stream of costs shown in Table 3.3 would be offset by approximately four years, resulting in the costs shown in Table C.17.[6] We further assume that annual maintenance and nonpersonnel-related reconstitution costs would be displaced by a similar interval.

- Larger numbers of production workers may be available when production of the next generation submarine begins. The construction of SSN-23 maintains the workforce for a time and provides a larger base to build upon when new production resumes. Our estimates of personnel released and remaining for different-year production starts are shown in Tables C.18 (including indirect support) and C.19 (production workers only).

- Though reconstitution costs (see Table C.20 and Figure C.17) for later restart years equal those at an inactive yard, workers coming off the SSN-23 line keep reconstitution costs at least as low in the interim as those for the overhaul case. Restarting production in FY01, for example, costs $1.7 to $2.0 billion in personnel-related reconstitution costs if the yard is inactive, but only $0.9 to $1.3 billion with SSN-23 work.[7] The effects of SSN-23 on postrestart schedule are analogous to those for cost (see Figure C.18). The pattern is also reflected in total reconstitution costs (Figure C.19).

Table C.17

**EB Shutdown Costs as a
Function of Next Start
(millions of FY92 dollars)**

Next Start	Total
FY95	2
FY96	12
FY97	29
FY98	44
FY99	50
FY00	58
FY01	83
FY02	87
FY03	90
FY04	134

---

[6]"Next start" and "restart" in this table and elsewhere in the discussion of this case refer to the start of construction on the NSSN (for consistency with the other shipyard analyses), not construction start for the SSN-23.

[7]Reconstitution costs for FY98 are a little higher than those for an inactive yard, because at that point SSN-23 is still occupying workers who could otherwise be available for the NSSN. The initial NSSN workforce is thus lower with SSN-23 than without it for a FY98 restart.

Table C.18

EB Personnel Released by Various Dates

Restart Year	Workers Remaining at Start of Year	Workers Released During Previous Year	Cumulative Workers Released
FY94	13400		
FY95	12900	500	500
FY96	11000	1900	2400
FY97	7600	3400	5800
FY98	4600	3000	8800
FY99	3300	1300	10100
FY00	1700	1600	11700
FY01 or later	300	1400	13100

Table C.19

Skilled Workforce Available When Production Resumes

Restart Year	Workers Remaining at Start of Year	Workers Released During Previous Year	Workers Available for Restart	Skilled Transfers This Year and After
FY95	10500	500	550	9450
FY96	8150	2350	2215	7335
FY97	5050	3100	3260	4545
FY98	3400	1650	2105	3060
FY99	2100	1300	1500	1890
FY00	1250	850	1025	1125
FY01	900	350	485	810
FY02	260	640	856	0
FY03	260	0	388	0
FY04	260	0	260	0

Table C.20

**Total EB Personnel
Reconstitution Costs
(millions of FY92 dollars)**

Next Start	Rate = 2	Rate = 3
FY98	351	633
FY99	524	828
FY00	708	1037
FY01	939	1328
FY02	1072	1470
FY03	1549	1880
FY04	1747	2125

NOTE: Case shown assumes fixed overhead of $150 million (pessimistic), early attrition rate of 5 to 10 percent (optimistic), and mentor:trainee ratio of 1:2 (intermediate).

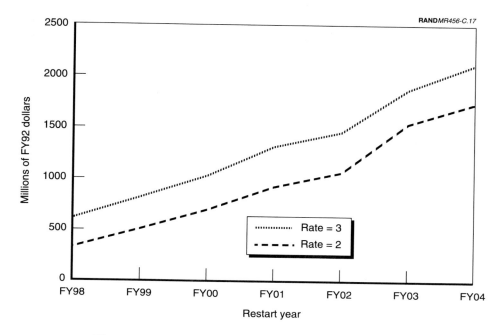

**Figure C.17—EB Personnel-Related Reconstitution Costs,
with SSN-23**

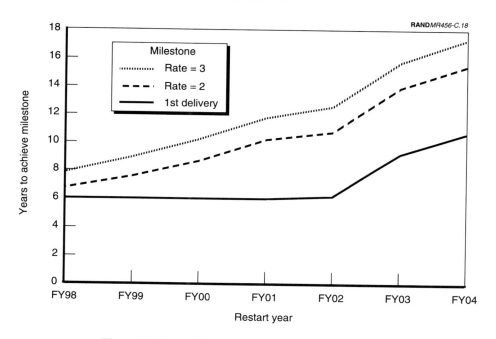

**Figure C.18—EB Postrestart Reconstitution Schedule,
with SSN-23**

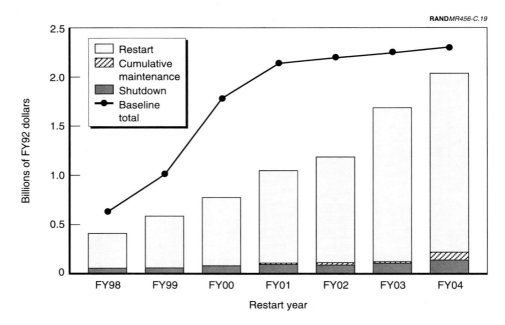

Figure C.19—EB Shipyard Reconstitution Costs, with SSN-23

SUMMARY

The shipyard costs of deferring production are presented in Table C.21 for all cases examined. Effects on current construction and the costs of shutdown and maintenance do not vary with target production rate. The costs of restarting production, and thus the total costs, do. Costs for target production rates of two and three per year are shown. Each cell in the table shows the costs incurred over all years for the category heading that row, given restart in the year heading that column.

Table C.21

Summary of Shipyard Costs of Deferring Production

Case and Cost Category	Cost ($M) for Restart in						
	FY98	FY99	FY00	FY01	FY02	FY03	FY04
Electric Boat, No Additional Work							
Current construction	40	60	80	80	80	80	80
Shutdown	78	88	134	134	134	134	134
Cumulative maintenance	16	24	72	120	168	216	264
Restart (2/yr)	337	709	1298	1747	1849	1869	1889
Restart (3/yr)	612	1038	1685	2054	2227	2247	2267
Total (2/yr)	471	881	1584	2081	2231	2299	2367
Total (3/yr)	746	1210	1971	2388	2609	2677	2745
Newport News, No Additional Work							
Current construction	0	0	0	0	0	0	0
Shutdown	68	68	68	68	68	68	68
Cumulative maintenance	61	107	153	199	245	291	337
Restart (2/yr)	478	832	1555	1869	1889	1889	1889
Restart (3/yr)	767	1181	1901	2247	2267	2267	2267
Total (2/yr)	607	1007	1776	2136	2202	2248	2294
Total (3/yr)	896	1356	2122	2514	2580	2626	2672
Newport News, with CVN-76							
Current construction	0	0	0	0	0	0	0
Shutdown	8	8	8	8	8	8	8
Cumulative maintenance	20	25	30	35	40	45	50
Restart (2/yr)	486	753	682	595	517	460	453
Restart (3/yr)	877	1026	939	852	771	742	763
Total (2/yr)	514	786	720	638	565	513	511
Total (3/yr)	905	1059	977	895	819	795	821
Electric Boat, with Overhaul Work							
Current construction	0	20	40	40	40	40	40
Shutdown	78	84	115	115	115	115	115
Cumulative maintenance	16	24	40	56	72	88	104
Restart (2/yr)	337	693	976	1048	1058	1068	1078
Restart (3/yr)	612	1023	1362	1430	1440	1450	1460
Total (2/yr)	431	821	1171	1259	1285	1311	1337
Total (3/yr)	706	1151	1557	1641	1667	1693	1719

Table C.21—continued

Case and	Cost ($M) for Restart in						
Cost Category	FY98	FY99	FY00	FY01	FY02	FY03	FY04
Newport News, with Overhaul Work							
Current construction	0	0	0	0	0	0	0
Shutdown	53	53	53	53	53	53	53
Cumulative mainte-nance	31	47	63	79	95	111	127
Restart (2/yr)	451	773	1023	1071	1081	1081	1081
Restart (3/yr)	740	1113	1412	1463	1473	1473	1473
Total (2/yr)	535	873	1139	1203	1229	1245	1261
Total (3/yr)	824	1213	1528	1595	1621	1637	1653
Electric Boat, with SSN-23							
Current construction	0	0	0	0	0	0	0
Shutdown	44	50	58	83	87	90	134
Cumulative mainte-nance	0	0	0	8	16	24	72
Restart (2/yr)	351	524	708	945	1078	1555	1809
Restart (3/yr)	633	828	1037	1334	1476	1886	2187
Total (2/yr)	395	574	766	1036	1181	1669	2015
Total (3/yr)	677	878	1095	1425	1579	2000	2393

NOTE: "Current construction" refers to effects on ships currently under construction. See text above and in Chapter Three for further explanation of categories and assumptions. Significant figures shown exaggerate precision of estimates to permit accounting for numbers of very different sizes.

BRITISH PRODUCTION RESTART EXPERIENCE

Great Britain has a long history of submarine construction and operates the world's third largest nuclear submarine force. Their nuclear submarine program began in the late 1950s and their first SSN, the *Dreadnought*, was launched in 1963 (see Figure D.1). The *Dreadnought* was a joint technology effort with a British-designed forward section and a U.S.-provided nuclear reactor (similar to the propulsion system of the USS Skate class). Currently, Great Britain's naval forces include some 12 nuclear attack submarines, several ballistic missile submarines, and a small number of both newer and older diesel-powered attack submarines. Structural and operating characteristics of British submarine classes are presented in Table D.1.

Great Britain is in the process of procuring a new class of ballistic missile submarines (Vanguard) to replace the Polaris-equipped Resolution class. They are also completing production of a fleet of four new diesel-powered submarines, the Upholder class. The last three of these boats are being built by Cammell-Laird, a yard that restarted conventional submarine production after a hiatus of almost two decades. This reconstitution of submarine production is of primary interest to our research. Appendix D examines the submarine production process in Great Britain and discusses the recent experiences at Cammell-Laird.

THE DESIGN AND PRODUCTION OF SUBMARINES IN GREAT BRITAIN

Organizations Involved in Design and Construction

The Sea System Controllerate of the Procurement Executive of the Ministry of Defence (MOD) develops the conceptual designs and manages the various submarine construction programs for the Royal Navy. Vickers Shipbuilding and Engineering Limited (VSEL) is the main submarine design and construction agent in Great Britain. Vickers's primary submarine construction facility is at Barrow-in-Furness, although they also build submarines at the Cammell-Laird yard, which became part of VSEL after the denationalization of British

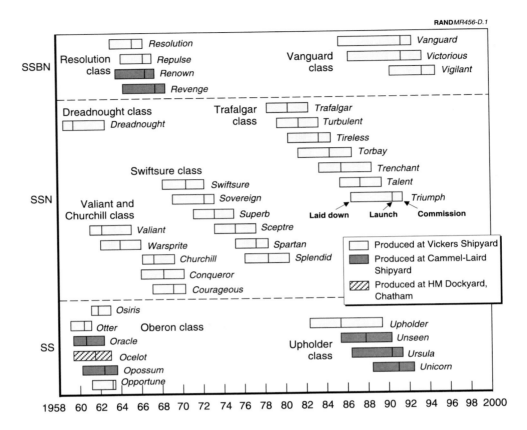

Figure D.1—Production and Commissioning History of British Nuclear Submarines

Table D.1

British Submarine Characteristics

Class	Number	Displacement[a] (tons)	Speed[a] (knots)	Length (ft)	Diameter (ft)	Crew Size
Vanguard	3[b]	15,000	−25	492	42	135
Resolution	4	8,500	−25	425	33	143
Dreadnought	1	4,000	−30	265	32	88
Trafalgar	7	5,208	−32	280	32	97
Swiftsure	6	4,900	−30	272	32	116
Valiant & Churchill	5	4,800	28	285	33	116
Upholder	4	2,455	20	230	25	47
Oberon	6	2,410	17	295	27	69

SOURCE: *Jane's Fighting Ships*, various years.

[a]Submerged.

[b]One to be authorized.

Shipbuilders in 1986.[1] In addition to submarines, VSEL designs and builds naval surface ships, manufactures various towed and self-propelled artillery, and has commercial ventures that include foreign military sales.

VSEL has not been capable of designing and delivering a complete operational system. The MOD contracts directly with a number of firms for the design and production of various weapons, fire control, and sonar subsystems on the submarines. The MOD provides these items to VSEL but has maintained responsibility for their integration. MOD intends to transfer this responsibility under a prime contract, after which it would retain design authority over the submarine and its major subsystems.

As there is effectively a sole supplier for submarines, the MOD relies on competition at the subcontract level to control prices. Vickers subcontracts for a number of submarine subsystems, including at times portions of the major hull structure. One of these subcontractors is Rolls Royce and Associates, who design and procure the naval nuclear propulsion plant.[2]

MOD insists that a certain proportion of the shipbuilding work be put up for such subcontracting even if the shipbuilder has the capability to do the work. Suppliers are not limited to British companies. Despite MOD's desire to maintain competition among subcontractors, the low historical production rate (roughly one per year) has necessarily resulted in several single-source arrangements for various subsystems. MOD has a core of cost accountants and cost estimators to assess the appropriateness of subcontractor costs. They judge whether anticipated costs should be reduced or completely disallowed. MOD's cost engineers use a variety of estimating techniques and historical contract data to estimate future costs. They avoid cost plus contracts and seek to agree on a firm price before a contract is awarded to ensure that the maximum incentive is placed on the contractor.

Organizations Involved in Overhaul and Refueling

Submarine overhauls, especially reactor refueling, take place at the specialized Royal dockyards. Nuclear overhaul is much more difficult than the nuclear aspects of new construction, in that it requires special facilities and skills to handle the irradiated fuel. VSEL looked into the possibility of bidding for refitting

---

[1]VSEL acquired Cammell-Laird in 1985.

[2]Rolls Royce and Associates (RRA) is owned by the Rolls Royce Industrial Power Group (IPG). RRA was created through the 1958 U.S./UK bilateral agreement for mutual cooperation in defense matters. The UK strictly adheres to the principles of this agreement in all nuclear issues. The defense nuclear work performed by RRA is rigidly separated from the commercial nuclear work performed by Rolls Royce IPG.

work but, having taken into account the investment required in training and nuclear safety certification, concluded that they could not compete with the Royal dockyards.[3]

The two Royal dockyards that have been used for submarine overhaul and refueling are Devonport in southwestern England and Rosyth in Scotland. These yards are owned by the government but operated by private companies. In June 1993, the MOD decided to direct all future submarine work to Devonport.

Production Rates and Budget Cuts

The one-per-year production rate of submarines has remained fairly stable over the past 30 years. Peaks in production activity have been most recently tied to the new class of ballistic missile submarines. Vickers deals with such peaks by subcontracting portions of the hull structure or by hiring temporary workers.

Great Britain, like the United States, is in a period of declining defense following the breakup of the Soviet Union and the relaxation of tensions in the world. As a result, production rates are expected to stretch out to one submarine every 18 or 24 months. Citing the need for efficient progress of work—fabrication, outfitting, and trials—and the need for a continuity of work to retain expertise, VSEL prefers that the production rate not drop below one submarine every 18 months.

Specific effects of budget cuts include the cancellation of the equivalent to the U.S. Seawolf program and a reduction in the buy of the Upholder class from an original ten to the current four. MOD officials are now designing and planning for a production run of a Batch 2 Trafalgar-class attack submarine. This new class involves minimum changes to the existing Trafalgar design. This is considered to be the most economic way of procuring additional submarines to maintain the desired force level.

The Barrow-in-Furness Yard and VSEL's Response to Varying Demand

In addition to building submarines, VSEL's Barrow-in-Furness yard has also built surface ships; for example, the aircraft carrier *Invincible* was constructed at Barrow. However, the recent award for the construction of a new helicopter carrier represents the first surface ship to be built at Barrow in over a decade.[4]

---

[3]The Barrow location presents certain problems for the handling of "dirty" fuel. VSEL estimates it would take seven years to attain the capability to refuel nuclear submarines at their Barrow facility.

[4]The hull will be built to commercial standards at a yard with tanker experience. It will then be transported to Vickers for outfitting. The resulting costs are expected to be approximately half of a military specification helicopter carrier.

(This contract award is being contested.) Employment at Barrow is currently approximately 7500, down from 14,500 in 1990. The workforce includes approximately 600 submarine design engineers and naval architects.

VSEL recently invested almost £250 million in a new production facility termed the Devonshire Dock Hall. This covered, modern facility contains accommodations for all necessary shops and support stores and was organized based on the principles of modular submarine construction. The facility also provides for the level launch of ships and submarines (the older method involved building on an inclined slipway). This new production facility is a key step in VSEL's drive for increased productivity and improved working conditions.

This facility was built because steady production was expected and available capacity was limited. Because of these limitations, Vanguard-class hull sections were built by other firms that did not have prior submarine construction experience. While these firms had the basic structural steel expertise, VSEL found they needed to provide management and technical expertise and placed engineers and managers on-site. Some difficulties occurred, but were manageable. Although these other firms had a higher cost (approximately 20 percent) and experienced schedule problems, they did suggest several useful process improvements. But because of the capacity limitations and the expectation of a class of 10 or 12 diesel submarines plus potential exports, submarine production at Cammell-Laird was restarted.

RESTARTING DIESEL SUBMARINE PRODUCTION

Building the Upholder class required meeting challenges both in the construction of Britain's first diesel submarine in 20 years and in the reinitiation of construction of submarines (of any kind) after a similar gap at Cammell-Laird.

Barrow-in-Furness

The MOD had decided on the construction of a new class of diesel submarines because their current diesel boats (Oberon class) were approaching the end of their service life and they felt there was a valid operational role for diesel submarines as a compliment to the nuclear boats. The lead boat in the Upholder class was designed and built by Vickers at Barrow-in-Furness.

The design stage was somewhat compressed and suffered from the lack of sufficient funds to adequately test various systems and construct mock-ups before

---

Swan Hunter, a British shipbuilder based at Wallsend on the River Tyne, announced it may be forced out of business after losing a bid to build this helicopter carrier for the Royal Navy.

they were placed on the lead boat. Several problems caused a delay in fielding the first of the class. The difficulties were associated with the main propulsion system and weapons handling and discharge. The general consensus was that the 20-year gap between diesel classes resulted in 20 years of technology "improvements" being built into the new class in one step. In addition, the difficulty and complexity of fitting all the modern systems, except the nuclear powerplant, into a smaller volume contributed to the delays. (The volume available in the Upholder class is some 50 percent of that for the Trafalgar class.) The original contract acceptance date for the lead boat was slipped by approximately two and a half years and cost increased by more than 25 percent.

The British thus ran into production problems for a similar but more technically sophisticated version of a type of ship that had been out of production for two decades.[5] This occurred even though submarines of the last diesel class were still operational and there remained a technical support staff familiar with the basic hull, diesel propulsion, and other subsystems of the earlier boats. The principal cause of the difficulties stemmed from insufficient testing and the lack of shore-based prototyping of the propulsion, weapons system, and other plant and equipment that had undergone improvements and technological advancements from the Oberon class. VSEL officials told us that if they had to do it over again, they would prototype and test all new systems, leaving at least a year between sea trials and launch of the second of the class.

Cammell-Laird

Cammell-Laird is a small shipyard with a long history of constructing naval ships. Although they built two Resolution-class SSBNs and two Oberon-class SSNs in the 1960s, they had constructed only surface ships (destroyers, frigates, and support ships) for the two decades preceding the start of the three Upholder-class diesel submarines. Cammell-Laird was restarting submarine construction after a hiatus of almost 20 years.

MOD requested competitive bids for the construction of the three follow-up Upholder-class boats in the mid-1980s. Vickers, Cammell-Laird, and a third shipbuilder submitted proposals. At the same time, naval shipbuilding in Great Britain was denationalized, with Cammell-Laird becoming a subsidiary of VSEL. Given the corporate relationship, VSEL withdrew their bid for the Upholder construction. Cammell-Laird was awarded the contract in January 1986,[6] be-

---

[5]This is typical of attempts to restart or duplicate advanced industrial facilities after a gap; see the discussion in Chapter Four.

[6]MOD and Cammell-Laird believed there would be a class of 12, plus additional foreign military sales.

fore Vickers had completed production of the first boat. While industrial base issues were considered, they did not have much effect on the decision, which was based almost entirely on price and value.

Since Cammell-Laird was "building to print," the turmoil in the design of the first boat caused some slippage in the acceptance date for the last three boats. Original contract build times varied from 4 to 4.5 years. Experience with the first ship demonstrated that the original schedule estimates were inadequate (see Table D.2). Therefore, the rest of the boats were allotted additional time. Given these agreed-upon delays, Cammell-Laird successfully completed the three boats on time and within budget. In fact, once some initial problems with deficiencies in documentation of the construction process were corrected, the quality of the boats built at Cammell-Laird was as good as that at Barrow. Interestingly, the level of weld repairs at Cammell-Laird for their first boat was less than the level at Barrow for the Upholder. This was attributed to the small pool of experienced welders at Cammell-Laird. As the construction program expanded to accommodate the second and third boats at Cammell-Laird, new personnel had to be hired and trained. This increase in production activity resulted in a dilution of the skill level of the welders and an increase in the number of weld repairs for the last two boats. That is, the expansion of the welding team to accommodate the increased production resulted in some degradation of performance and quality.[7] The successful cost and schedule achivements, however, rested upon careful planning, the good fortune of being able to draw on VSEL expertise, and considerable effort to apply that expertise through training and consulting.

Table D.2

Upholder-Class Submarine Planned vs. Original Schedules

Submarine Class	Builder	Contract Award Date	Original Planned Delivery	Actual Delivery	Difference (months)
Upholder	VSEL	Nov 1983	Jun 1988	6 Dec 1990	30
Unseen	CLS	Jan 1986	Feb 1990	July 1991	16
Ursula	CLS	Jan 1986	Oct 1991	May 1992	6

NOTE: Unseen and Ursula were delivered on time according to the revised schedule.

---

[7]A decrease in quality resulting from an expansion of the production program was also experienced in the United States during the construction of the Los Angeles-class submarines and by several public shipyards when overhaul workloads were significantly increased.

Given their corporate relationship and the fact that Cammell-Laird had not built a submarine in almost 20 years, VSEL provided significant managerial and technical help to Cammell-Laird and assisted in the vendor relationships. Some of the hull rings for the first boat at Cammell-Laird were built at Barrow by VSEL. Cammell-Laird had a workforce experienced in building ships to military specifications and this workforce formed an adequate base for the submarine construction. They also had a production facility that was well suited for building submarines.

The reconstitution of Cammell-Laird to build submarines included multiple training courses, establishment of welder qualifications, the reconstitution of the quality control department, and a redefinition of all quality procedures. A good deal of on-the-job training was provided by VSEL experts, including bringing Cammell-Laird people to the Barrow facility to gain experience. During the first year, VSEL technical management and senior trade people conducted the program while Cammell-Laird staff gained experience. VSEL also transferred production tooling to the Cammell-Laird facility. All together, it was nine to twelve months before production began at Cammell-Laird. Over the course of constructing the three submarines, the workforce at Cammell-Laird increased from 1000 to approximately 2500.

The end result was a successful program that met revised schedule and budget goals. However, the compression of the shipbuilding industry in Great Britain has adversely affected Cammell-Laird. The shipyard closed after the delivery of the third Upholder-class submarine in 1993. Currently, VSEL is trying to sell the Cammell-Laird shipyard. They have invested approximately £50 million, which will cover severance pay and expected losses in book value when the shipyard is sold. The plan calls for the yard to be mothballed for about 12 months, to preserve the yard's shipbuilding capability for the next year. The company is managing the shutdown to minimize the impact on its employees. The reduction of force has been carried out in phases, and the company played an active role in retraining and relocating those being laid off.

Conclusion

The experience at Cammell-Laird suggests that submarine production can be reconstituted after a hiatus of several years *if core knowledge and personnel with ongoing submarine engineering and production experience are available and if the yard to be reconstituted has an adequate base of personnel skilled in military ship construction. Clearly, the key element of this reconstitution was the transfusion of managerial and technical expertise from Vickers.* Some hiring and training are certainly necessary, but the schedule and cost implications are manageable.

In reflecting on the British experience when considering an extended gap in U.S. submarine production, it is important to keep in mind this availability of continuous submarine production experience at another yard in the same corporate organization. It is also important to keep in mind that the submarines built at Cammell-Laird were diesel, not nuclear. The reconstitution of nuclear capabilities, including licenses, facilities, and skilled personnel would have made the problem much more difficult. It would have taken more time and been more expensive. VSEL and MOD contend that Cammell-Laird would not have been considered if the Upholder class were nuclear-powered.

FRENCH PRODUCTION EXPERIENCE

France has a long history of building and operating submarines. Currently, the French fleet includes five nuclear ballistic submarines (Le Redoutable and L'Inflexible classes), with two more under construction (Le Triomphant class), six nuclear attack boats (Rubis class), and several older diesel attack submarines (Agosta and Daphne classes). Figure E.1 shows the production and commissioning history of the last five French submarine classes. Table E.1 identifies nuclear submarine characteristics.

ORGANIZATIONAL STRUCTURE

The General Delegation for Armament (DGA), under the French Ministry of Defense, is the main agency for the design, production, and repair of military weapon systems. The Directorate for Naval Construction (DCN) is the organization within DGA responsible for naval ships, both submarines and surface ships. DCN maintains a core of design engineers at their Paris headquarters and operates several shipyards throughout France. These shipyards are nationalized installations that build and maintain various types of ships and shipboard systems and act as home port for various ships in the French Navy.

For example, DCN Cherbourg provides the full-scale design and construction for all submarines in the French Navy. Most submarine repairs, including major overhauls and reactor refuelings, take place at Brest and Toulon, the home ports for the French Atlantic and Mediterranean fleets, respectively. Other DCN shipyards or state-owned organizations are responsible for various submarine systems; for example, DCN Indret designs and builds the pressure vessels for nuclear reactors and Technicatome designs and builds the nuclear reactors.

THE DESIGN AGENT: DCN PARIS

The Paris headquarters of DCN houses the overall management functions and the conceptual design teams. The French submarine technical community is

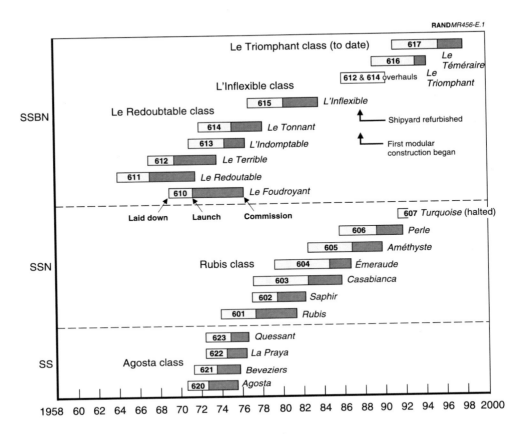

Figure E.1—Production and Commissioning History of the Last Five French Submarine Classes

Table E.1

French Submarine Characteristics

Class	Number of Submarines	Displacement (tons)	Length (ft)	Diameter (ft)	Crew Size[a]
Rubis	7[b]	2,640	242	25	65+
Le Redoubtable	5[c]	8,045	422	34.8	114+
L'Inflexible	1	8,080	422	34.8	127+
Le Triomphant	4[d]	12,700	453	41	110+

[a]French nuclear submarines are manned with two crews to allow them to remain at sea for the maximum amount of time.

[b]An eighth was canceled.

[c]Le Redoutable was the first French nuclear-powered ballistic submarine.

[d]First two are under construction, a third ordered, one or two more are to follow (original plan was for six).

relatively small. Teams typically remain together throughout the life of a submarine class, although they are located at multiple sites. Personnel are routinely rotated through the design activities in Paris and Cherbourg, submarine maintainance facilities in Brest, and research and development in Toulon. DCN in Paris makes the conceptual and preliminary designs, and Paris maintains the overall design responsibility. For full-scale detailed design and production design, the team moves to Cherbourg, where it is then responsible for the engineering design of any major modifications or modernizations that occur during the life of the submarine.

Although the French program is small in terms of number of submarines, the French believe that the continuous design and production of submarines is necessary to reduce overall costs and schedule delays and to maintain the quality of their fleet. Their approach is to stagger the designs of their ballistic missile and attack submarines and to fill the "gaps" in new class designs with engineering work on major overhauls and modernization of existing boats. This scheme is diagrammed in Figure E.2.

By staggering its design efforts, DCN maintains continuity in its design teams and avoids peaks and valleys in funding requirements. It believes continuity in

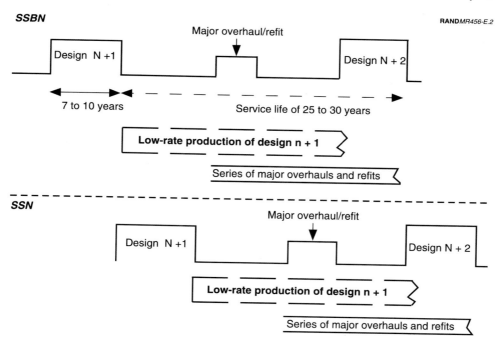

Figure E.2—The French Stagger Submarine Design Demand

actual practical designs is necessary to attract young engineers into the submarine field and to keep the best and brightest of those engineers. (This lends support to Appendix A's conclusions regarding continuity of design expertise.)

Because the French build only a few of each design and produce at a very low rate (one boat every two or three years), there is typically only one supplier for the various submarine components. Despite the lack of competition, DCN believes that low-rate production is necessary and preferable to stopping production for any period of time.

DCN is actively pursuing cooperative agreements with other countries such as the United States and Great Britain. France, like its allies, is facing dramatic cuts in its defense budgets, and DCN feels that cooperative agreements and partnerships are necessary to maintain French expertise in the submarine field in the future world of limited funding.

THE SHIPYARD: DCN CHERBOURG

DCN Cherbourg has over a century of experience in designing and building submarines and is the sole French shipyard for submarine construction.[1] It is also the home port for several smaller ships in the French Navy and provides repair and maintenance for those ships. It has approximately 4500 full-time employees and hires local "temporary" subcontractors based on the workload. At Cherbourg, approximately 730 people are involved in submarine design, including some 130 engineers and 600 technicians.

Although Cherbourg is a very old shipyard with some facilities dating from the days of Napoleon, the French have recently invested nearly 3 billion francs in new production facilities that use modern construction techniques. The facilities and the permanent workforce are sized to produce one submarine every two years. The facilities have the capability to surge to higher production rates, although additional temporary hires would be necessary to maintain higher rates. Currently, the lead boat in the Triomphant class of SSBNs is nearing completion and the second boat, *Le Téméraire*, is part way through the production cycle. The third submarine has been ordered and construction of one or two additional boats in this class is planned.

---

[1]DCN Brest was also involved in submarine construction earlier in this century. Brest built its last submarine (Daphne class) in the late sixties.

Budget reductions have resulted in the cancellation of two Rubis-class attack submarines[2] and a reduction in the number of new SSBNs from the original six to the current four. The production cycle for the four SSBNs has also been lengthened from one boat every two years to one boat every three years. This level of submarine production is not sufficient to keep personnel fully employed, so the shipyard actively seeks other work to fill the gaps and unevenness in submarine production workloads. This work includes the maintenance and overhaul of the several ships (none of them submarines) homeported at Cherbourg and the construction of portions of the new frigates being built at another DCN shipyard.

SSBN "RESTART": THE *LE TRIOMPHANT* EXPERIENCE

There was about a four-year gap between commissioning of *L'Inflexible* and the start of *Le Triomphant* in June of 1989.[3] During this gap in SSBN production, DCN Cherbourg embarked on a major modernization of two of the Redoutable class of SSBNs. This *refonte*[4] was very unusual and very extensive and resulted in a workload over six years that equaled the construction of a new boat. These overhauls helped maintain the submarine production skills and kept the shipyard employees productive. They also helped somewhat to exercise engineering design skills.

The overhaul of the two SSBNs and the continued construction of the Rubis-class attack submarines helped lessen the negative impact of low production when the new class of SSBNs was started. There were some problems with system integration and physical arrangement, getting used to new facilities, retraining new workers, and with communications between the various subsystem design groups, but these did not adversely affect the construction process.

---

[2]The seventh boat in the Rubis class, the *Turquoise*, was halted after the pressure hull and reactor components were completed; this boat is currently in storage. The eighth boat, the *Diamant*, was canceled.

[3]This is not, of course, a restart in the sense we use it in the rest of this report, and it is not unique. (Electric Boat "restarted" SSBN production in 1974, seven years after the last Franklin-class submarine was commissioned.) However, we felt that the way the French dealt with this lull in demand for new construction was interesting and potentially applicable to a full stop in submarine production.

[4]The *refonte* is a mid-life modernization that can last up to three years. A French submarine has two other types of repair or overhaul during its life. One occurs approximately every three months (or when a submarine returns from sea duty) and lasts for three to five weeks. The other occurs every five years, takes about a year to accomplish, and can include a reactor refueling.

FUTURE OUTLOOK

With work on the two additional planned SSBNs, DCN Cherbourg managers feel they can maintain the production skills of their workforce with their current submarine program and with the inclusion of other production and overhaul work. They also feel they can increase the production rate if needed by hiring additional personnel from the local pool. These temporary hires usually have a background in the desired skills. Furthermore, Cherbourg provides several weeks of training to enhance the basic skills and the application of those skills in the submarine construction area.

CONCLUSION

Although faced with very low production rates, DCN has decided that continuous production is preferred to stopping and starting the production lines. It has consolidated all submarine production capability in one shipyard (although several other shipyards work on subsystems) and has reduced its vendor support to basically one contractor in each of the various systems. When necessary, the shipyards that perform overhauls are called upon for new construction activities. The viability of the French commercial nuclear industry helps maintain nuclear capabilities for submarines. The French believe that submarine production skills can be maintained through overhaul and repair work and that new hires can be trained in submarine production skills given a basic competency when hired.

The bigger concern for DCN is the maintenance of design skills—a concern shared by U.S. submarine builders. The leadership of DCN believes that, to maintain a competence in submarine R&D, design and engineering, and new construction, a constant flow of new work must be maintained in the yards. Expertise at universities and in paper studies is *not* sufficient. Concern was also expressed by DCN officials about sending negative signals to the next generation of potential submarine designers if submarine production is allowed to stop for a period of time. With so few boats produced and a large gap between new classes of submarines, DCN must try hard to keep its design engineers actively involved and interested in new submarine designs. DCN is currently facing a gap in submarine design, and is actively seeking ways to fill that gap. Possibilities include the engineering work associated with the modernization of the Rubis-class submarines, mid-life modernization of the Triomphant class, design of a new class of attack submarines, and cooperative agreements with allies such as the United States and Great Britain.

WORKFORCE RECONSTITUTION MODEL AND GENERIC RESULTS

Appendix F expands on Chapter Three's analysis of rebuilding a production workforce.[1] Here, we describe in detail the model whose results are given in that chapter. We also elaborate on our analysis of factors influencing the schedule and cost consequences of rebuilding a production workforce in an industry where production work is phasing down toward a hiatus or has already stopped. The analysis considers construction of a new attack submarine (NSSN) that is not expected to involve significantly different methods than those used in construction of current submarines (Trident, Los Angeles, and Seawolf classes). However, the approach and some of the results should be applicable to construction of other types of ships and even to other industries.

The schedule and costs for the early submarines that would be produced after a decline in workforce depend on several factors, including the size of the workforce at restart, the rate of hiring, the experience level of new hires, the costs to hire and train, and the amount of fixed overhead costs. Because there is a significant degree of uncertainty regarding what values these factors would assume and how important they are relative to each other, we wanted to be able to quantify schedule and cost effects over a range of values. It was for that reason that we constructed an analytic model.

The model begins with an initial workforce that evolves toward a steady state through hiring and attrition. The steady state is defined in terms of a predetermined annual production rate. If the initial workforce is smaller than the steady-state workforce required to produce submarines at the predetermined rate, then additional personnel are hired. If there are other construction programs in the shipyard that are winding down, then workers released from these programs are available to the restart program.

---

[1]We consider only production workers. Supervisors, foremen, engineers, and indirect personnel are not explicitly included, but wrap rates account for their costs.

In the model, *the workforce is not differentiated by occupational skill.* In the real world, shipyard managers must deal with all the factors included in this model for each category of workers—welders, riggers, electricians, etc. This implies a source of inefficiencies not addressed in the model. More important, skill mix differs between the start and end of submarine construction, but, in projecting workforce buildup, the model cannot take that into account. Thus, where workforce drawdown profiles from Chapter Three and Appendix C suggest an initial workforce of 1000 or 2000, these are mid- or end-phase workers coming off submarine lines that are winding down. Initial-phase workers have since dispersed. Because of this, the model results should be considered, particularly in the early years following restart, as conservative estimates of the consequences of the various factors.

To determine effects on schedule, submarines are counted as delivered when sufficient man-hours of work have been completed. As for cost, we input the cost of a "typical" (or "steady-state") submarine, which the model then augments. The early submarine costs will be higher than the costs of "steady-state" submarines because of higher hiring and training costs; the relative inefficiency of the rebuilding workforce, which is a result of larger numbers of unskilled workers; and amortization of shipyard fixed overhead costs over fewer submarines during the rebuilding period.

This appendix is divided into two parts. First, we describe the parts of the model that relate to workforce and workload analysis and the generic (non-shipyard-specific) schedule results yielded. Then, we turn to the estimation of costs and the generic results obtained from that effort.

SCHEDULE

In modeling workforce buildup and workload distribution, our objective was to quantify the effect of factors such as initial workforce size and attrition on schedule variables—that is, the time to produce the first ship and time to reach the target annual production level. We discuss the methodology first, then the generic results.

Approach

Modeling the reconstitution of the workforce and its production entails answering two questions: How fast can the workforce be rebuilt in terms of persons and person-hours per year and how do those hours translate into submarine completions in a way that allows for progress toward a stable, sustained rate of production?

Rebuilding the Workforce. The model is set up to represent the growth of an initial workforce to some target level. For our generic analyses, a range of initial workforces is used. For specific shipyard analyses, the initial workforce is determined from the numbers of workers laid off in recent years. We assume that, for restart in a given year, 90 percent of the workers released the previous year can be rehired, along with 20 percent of those released the year before that. (We understand from the shipyards that these are optimistic estimates, so our reconstitution cost estimates are conservative on this account also.) The model also accommodates transfers from other shipyard work as workforce levels decline. These transfers occur each year until any existing work is totally phased out.

The target workforce level is determined by the steady-state (target) production rate, the steady-state number of worker-hours per submarine, and the number of chargeable hours per worker-year.[2] Using 1760 chargeable hours per worker-year and 10M (where M = million) hours per submarine, the workforce sizes for steady-state production of one, two, and three submarines per year are 5682, 11,364, and 17,045, respectively.

In the model, the workforce is distributed across 31 experience categories, corresponding to 0 through 30 years. Each experience category is characterized by (1) a relative efficiency, (2) a relative compensation rate, (3) attrition rates (the percentage leaving the workforce each year), and (4) the initial number of personnel. The initial workforce is assumed to have no workers with fewer than five years of experience or more than 25 years of experience.[3] This assumption reflects the expected conditions in a declining industry, in which there has been no hiring of entry-level workers for a few years and workforce reduction incentives have induced senior personnel to retire.

The relative efficiencies capture the inefficiency of a workforce that is undergoing a rebuilding process by hiring large numbers of unskilled personnel as compared to a steady-state workforce that has significantly fewer unskilled workers. Suppose that an unskilled worker can only do half the productive work of a fully skilled worker. In the model the unskilled worker would be characterized by a relative efficiency of 1/2 and would be credited with 1760/2 hours of productive work per year. Workers at the lowest level would have the lowest efficiency and the efficiency values would increase with experience, reaching a

---

[2]The number of chargeable hours reflects those hours that workers charge to contract work. It excludes vacations, holidays, sick days, jury duty, personal business, and any other charges that accrue against overhead or fringe benefits.

[3]Several specific initial-workforce experience distributions were tested to determine their impact on various workforce characteristics, to ensure that the results presented here are not significantly influenced by the choice of starting conditions.

maximum of one. Real-world data regarding relative efficiencies as a function of years of experience are not readily available. For this study, we set the work-force relative efficiencies by years of experience as shown in Table F.1. These hypothetical values were reviewed by submarine industry personnel and judged to be reasonable.

The relative compensation rates are based on representative submarine ship-yard pay scales (see Table F.2).

There are two sets of attrition rates, one that applies during the restart and another that applies to the steady state. Steady-state attrition rates are given in Table F.3. Although the shipyards do not have attrition data by experience

Table F.1

**Assumed Relative Efficiencies
by Experience Level**

Years of Experience	Relative Efficiency
0	0.40
1	0.50
2	0.60
3	0.70
4	0.80
5	0.85
6	0.90
7	0.92
8	0.93
9	0.94
10	0.95
11	0.96
12	0.97
13	0.98
14	0.99
15+	1.00

Table F.2

**Assumed Relative Compensation
Rates by Experience Level**

Years of Experience	Relative Compensation
0–4	0.6
5	0.7
6	0.8
7+	1.0

Table F.3

**Assumed Steady-State Attrition
Rates by Experience Level**

Years of Experience	Annual Attrition Rate (%)
0–8	5
9–12	4
13–16	3
17–20	2
21–24	1
25–28	0.5
29	0
30	100

level, these values seem reasonable for two reasons. First, they decrease with experience, which is generally believed to be true. Second, they result in a steady-state distribution of headcount by years of experience that is reasonably close to the actual recent distribution in the shipyards, where the workforce has matured. In the model, attrition is removed before the workers at each level move up in experience. (The model works in annual steps.) The attrition rate at the 30-year level is set to 100 percent to prevent accumulation of a top-heavy workforce.

Several alternative early attrition rates are discussed below. These early rates operate for the first five years of a restart.

An important characteristic of the model dynamics is that new workers can be hired only at the lowest apprentice level (zero experience), and the maximum number of new hires each year is determined by the number of workers having at least five years of experience and by a mentor: trainee ratio. This ensures that there is some specified number of skilled workers to serve as mentors to apprentice-level workers. Semiskilled workers (those with two through four years of experience) cannot serve as mentors, but do not require individual mentoring. Mentoring ratios between 1:1 and 1:5 were examined.

The elements of the model described to this point permit analysis of the workforce buildup, and the total and effective worker-hours by year. The workforce buildup is simply the growth in size of the workforce each year as the result of hiring and attrition. After some number of years, the workforce reaches the target level for steady-state production, beyond which the distribution across experience levels evolves to a steady-state condition.

The total workforce hours each year is the number of workers times the number of chargeable hours (1760) per worker-year. If all workers were in the highest skilled brackets, then the amount of effective, or productive, work would be the same as the chargeable hours. With a mix of workers across all levels of experience, the number of effective hours will be less than the number of chargeable hours. The rebuilding workforce takes longer than the steady-state workforce to accumulate the number of effective hours necessary to complete a submarine because of the larger portion of unskilled workers and the smaller size of the initial workforce.[4]

Completing Submarines. The steady-state workforce is sized to start and complete one, two, or three submarines per year. Of course, for any given year, the ones started are different from the ones finished, because it takes (nominally) six years to build a submarine. Thus a shipyard sized to produce (on average) one submarine per year would have six submarines in various stages of construction in the yard at any given time (on average). For simplicity, we assume that the work required to build a submarine can be equally divided over the six years. Then, a "balanced" shipyard—one with a stable "age distribution" of submarines under construction—would at the end of a given year have one submarine 1/6 complete, another 2/6 complete, and so on, with the last ready to deliver. Summing these values indicates that a "balanced" yard has 3.5 equivalent submarines at the end of the year, one in completed form and 2.5 as work in process (WIP). Similarly, at the end of each six-month period, a "balanced," two-submarines-per-year yard will have one complete and 5.5 equivalent WIP submarines; and at the end of each four-month period, a "balanced," three-submarines-per-year yard will have one complete and 8.5 equivalent WIP submarines.

The model does not explicitly distribute the work output across submarines. Instead, it counts deliveries if at year's end enough effective worker-hours have been accumulated to reach or exceed the end-of-year balanced-yard WIP. For example, in a one-submarine-per-year yard, enough worker hours to build 3.5 submarines must have been accumulated (leaving 2.5 WIP subs after delivery). This rule can be used to determine steady-state deliveries, but it can present problems when used to determine early deliveries. In cases where the initial workforce is a significant percentage of the steady-state workforce, the model will accumulate sufficient WIP to indicate submarine deliveries in less than six years. This is not consistent with current submarine construction experience and six years is used as the minimum.

---

[4]The model is based on a nominal submarine that requires a constant number of effective hours to produce. The analyses presented here are based on a submarine to whose construction a steady-state workforce would charge 10 million hours. For the cases examined in this study, the corresponding number of effective worker-hours is in the neighborhood of 8.8 million. A workforce short of steady state would charge more than 10 million hours to build the same 8.8-million-effective-hour submarine.

A key question in rebuilding a submarine construction workforce is, when can that workforce sustain a steady-state production rate that will maintain the desired fleet size? Figure F.1 shows an example of the profile of WIP and deliveries for restart of a three-ships-per-year yard using the rule described above. Deliveries see-saw between zero and one per year from years 6 through 14, then between two and three for years 15 through 18. In year 19, the sustained rate of three is reached. Examination of many such model-generated initial delivery patterns resulted in the judgment that a steady-state rate of three ships per year is achieved with delivery of the eighth submarine. Similarly, steady state for a two-ships-per-year yard is achieved with the fifth delivery, and the second delivery constitutes rate achievement for a one-ship-per-year yard.[5]

Generic Results

We now present model results indicating the effects of variations in initial workforce size, attrition, and mentoring ratio on postrestart schedule, for

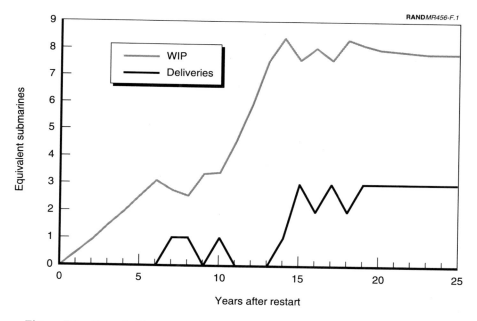

Figure F.1—Sample Postrestart Profile of Work in Progress and Deliveries, for Sustained Production Rate of Three per Year

---

[5]Note that WIP in Figure F.1 stabilizes around 8. If WIP has reached 8.5 or more at year's end, enough submarines are counted as delivered to bring the WIP below 8.5. Thus, end-of-year WIP could run anywhere from 7.5 to 8.4, with numbers near 8 being typical.

maximum production rates of one, two, and three submarines per year. We report the two schedule variables mentioned above—time to first delivery and time to reach sustained production rate. The first is important because, when applied to a specific shipyard scenario, it indicates how far in advance of a required delivery date production must be restarted and thus limits the length of a production gap. The second is important because it indicates when a shipyard is able to maintain a given force level under steady-state conditions (e.g., a force level of 60 submarines when the sustained rate is two ships per year with a service life of 30 years).

As the size of the initial workforce increases, the time required to deliver the first submarine and to reach rate decrease (see Figures F.2, F.3, and F.4[6]). Starting with only 250 skilled workers, a rebuilding shipyard will require about 12 years to reach a sustained production rate of one ship per year, about 15.5 years to reach two per year, and about 17.5 to reach three per year. If the initial workforce is instead 2000, only about half as much time is required to reach the sustained rate. For production rates of two or three, however, even these times are roughly two to four years longer than those required for first delivery (the minimum six for workforces of 1000 or more).[7]

We tested three sets of early attrition rates. For the cases shown above (and for the results in Chapter Three and Appendix C), early attrition rates were set low (see Table F.4); they represent the most optimistic values for a rebuilding shipyard. The most pessimistic set was necessary for the model to produce total workforce average attrition rates of nearly 30 percent per year, which were experienced by Electric Boat in the early to middle 1970s when EB was expanding their workforce for the early Trident- and Los Angeles-class construction.

The early attrition rate does not have a great deal of influence on schedule or cost (see Figure F.5). For an initial workforce of 250 (shown), going from low to medium attrition adds about one year to the time to reach sustained rate, and going from medium to high adds between one and two years. For larger initial workforces, the sensitivity is reduced.

---

[6]Note that time to reach first delivery is independent of the target production rate. We show it for all three rates to ease comparison with time to reach rate.

[7]Note that although workforces of 250 and 1000 approximate the residual cadre and overhaul workforces for cases reported in Chapter Three and Appendix C, these results are different because in the specific shipyard scenarios, transfers are available.

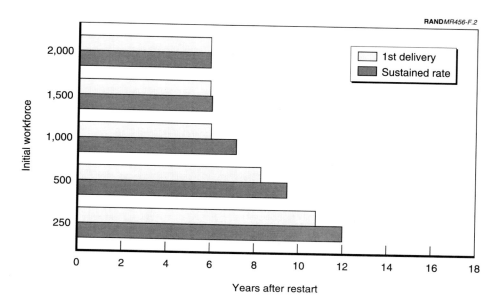

**Figure F.2—Time to Achieve First Delivery and a
Sustained Production Rate of One per Year**

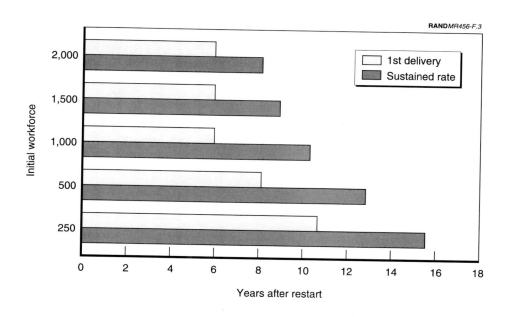

**Figure F.3—Time to Achieve First Delivery and a
Sustained Production Rate of Two per Year**

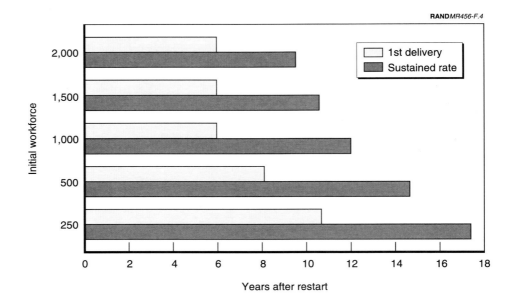

**Figure F.4—Time to Achieve First Delivery and a
Sustained Production Rate of Three per Year**

The mentor:trainee ratio is much more influential. In results presented so far, the ratio has been one skilled worker for two trainees. As shown in Figure F.6, the time to achieve a sustained rate of three ships per year is halved going from a mentor:trainee ratio of 1:1 to 1:5.[8] The largest change occurs between rates of 1:1 and 1:2. The effect is not quite as large for lower steady-state rates. The re-

Table F.4

Tested Early Attrition Rates, by Experience Level

Years of Experience	Attrition Rate (percent)		
	Low	Medium	High
0	10	25	50
1	7.5	20	40
2	5	25	30
3	5	10	20
4	5	7.5	10

---

[8]The 1:1 ratio was obtained from a public shipyard. The smaller numbers reflect the possibility that in time of need, such ratios are likely to be permitted.

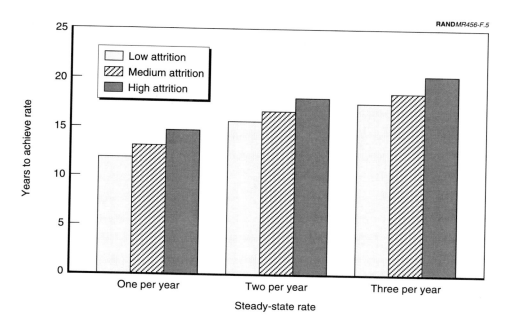

**Figure F.5—Time to Achieve Rate Is Not Very Sensitive to
Level of Early Workforce Attrition**

sults are for an initial workforce of 250 and for low early attrition. The sensitivity is slightly reduced for larger initial workforces and for higher initial attrition rates.

COST

The measure of personnel-related reconstitution costs used in this study is how much more the submarines constructed during the workforce rebuilding period cost than they would if they were produced at steady state. Because of the rebuilding process, the early submarines will cost more than the steady-state submarines to produce. This extra cost will dampen out with time as the delivery rate builds up and the workforce matures into a steady-state distribution of experience levels. We now describe how we estimated the reconstitution cost penalty and show how it is influenced by the factors described above.

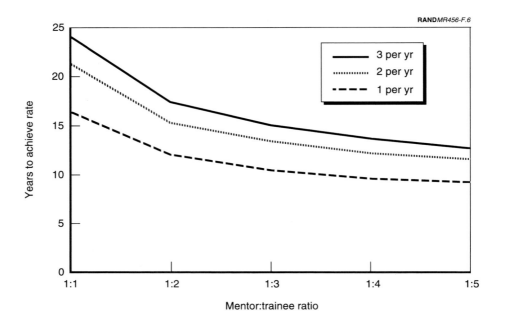

**Figure F.6—Time to Achieve Sustained Rate Is Sensitive
to Mentor:Trainee Ratio**

Approach

To determine the cost penalty, the model takes explicit account of the labor force buildup, in terms of efficiencies and compensation rates; the costs of hiring and training, which will be disproportionately high during a restart; and the allocation of fixed overhead costs, which will be spread over fewer deliveries during the buildup. Labor efficiencies and compensation rates are built into the model in the form of values for each level of experience, as already described. Treatment of the other items is described in the following paragraphs.

Because the model does not consider each individual worker skill category, representative average hiring and training costs are used. A $5000 cost is assessed for each hire. The lowest five levels (0–4 years experience) are considered apprentice/semiskilled workers. Workers at these levels receive yearly training, and an annual training cost of $3000 is assessed. A worker who does not leave before completing the training will accrue a total of $15,000 of training costs.

Workers who leave because of attrition incur training costs of $3000 per year for those years they are in the program.[9]

The higher fixed overhead is a major part of the cost penalty, so the distribution of overhead between fixed and variable components is important. Because the distribution between these components for future submarine production cannot be known precisely, we analyzed values of $50M, $100M and $150M of fixed shipyard overhead. These values span the range of expected distribution between fixed and variable costs.

For our nominal (steady-state) submarine costing $600 million when produced at three per year, we assume that $150 million is for overhead accruing to that ship (this is *not* the shipyard overhead figure just mentioned) and the other $450 million is for direct labor and material.[10] The submarine will bear one-third of the fixed shipyard overhead; if the latter is $150 million, the submarine would thus be charged $50 million. The other $100 million of overhead accruing to that submarine would be variable.[11]

In Table F.5, we show the distribution of costs for the case just described and for others with lower production rates or shipyard overhead. The case described is at the lower right in the table. Lowering the production rate—moving to the left across the table—will increase the fixed overhead allocated to each submarine but should not change the variable overhead. Reducing the fixed shipyard overhead—moving up the third column—naturally reduces proportionally the amount borne by each ship and raises the variable overhead to reach the $150 million per ship total assumed for three ships per year. The effect of moving from the high fixed shipyard overhead to the low value is to quadruple the variable: fixed ratio for each production rate.

As stated above, the purpose of the model is to analyze the inefficiencies associated with rebuilding a workforce. Consequently, we assume that the steady-state direct labor and material costs are the same, regardless of production rate or total quantity produced. "Learning" or "cost improvement," excepting what is inherent in the rebuilding of the workforce, is ignored. (Shipyard experience indicates that submarine learning curves are fairly flat.)

---

[9]The model does not, of course, track workers individually. At any given end-of-year step, the appropriate attrition rate is applied to the number of workers in each experience cohort and the number of workers remaining is multiplied by $3000 if the cohort now has 1–4 years of experience.

[10]$300 million is for labor and $150 million is for material. The latter covers shipbuilder material and does not include government-furnished equipment such as the reactor, combat systems, etc. The $600 million shipyard cost may be between 50 and 70 percent of the total submarine cost, depending on the cost of the government-furnished equipment.

[11]If the distribution of cost between direct and overhead does not look quite right to readers unfamiliar with shipyard accounting, the direct labor cost ($300 million here) includes certain items such as fringe benefits that are considered overhead in other industries.

Table F.5

**Distributions of Fixed and Variable Overhead Costs Examined
(millions of FY92 dollars per submarine)**

Cost Element	Peak Production Rate		
	1 per yr	2 per yr	3 per yr
Low fixed overhead			
Direct labor & material	450	450	450
Variable overhead	133	133	133
Fixed overhead	50	25	17
Total cost	633	608	600
Medium fixed overhead			
Direct labor & material	450	450	450
Variable overhead	117	117	117
Fixed overhead	100	50	33
Total cost	667	617	600
High fixed overhead			
Direct labor & material	450	450	450
Variable overhead	100	100	100
Fixed overhead	150	75	50
Total cost	700	625	600

To determine the cost penalty associated with workforce reconstitution, the model calculates the number of effective worker-hours available from the workforce each year. Dividing this by the number of effective worker-hours per steady-state submarine results in the number of equivalent submarines produced each year. Multiplying the number of equivalent submarines by the steady-state submarine cost gives the cost for constructing those submarines with a steady-state workforce. Subtracting this cost from the corresponding annual cost determined by the model for the rebuilding workforce yields the annual cost penalty for restarting. Summing this over the period from restart to achievement of steady state yields the total cost penalty.

The annual cost penalties are highest at the start of the rebuilding process. Because hiring is restricted to zero experience apprentices, there are cohorts of workers that work their way through the experience levels in the model and there are corresponding fluctuations in the overall workforce efficiency. Thus, the annual cost penalties fluctuate with the changing composition of the rebuilding workforce while generally decreasing as the workforce builds up and improves in overall efficiency. These fluctuations dampen out and the cumulative cost penalty approaches a steady-state value.

Generic Results

Figures F.7 through F.9 display the total reconstitution cost penalty for shipyards with sustained production rates of one, two, and three ships per year. The

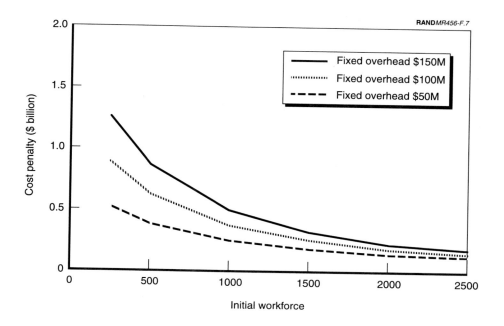

**Figure F.7—Workforce-Rebuilding Cost Penalty for a Sustained Production
Rate of One Ship per Year**

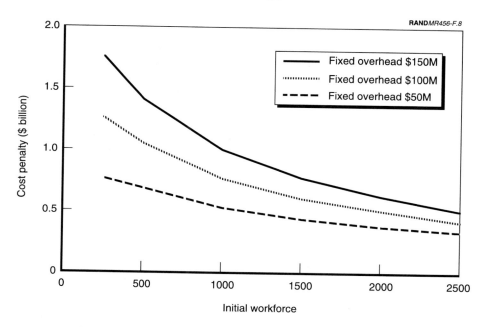

**Figure F.8—Workforce-Rebuilding Cost Penalty for a Sustained Production
Rate of Two Ships per Year**

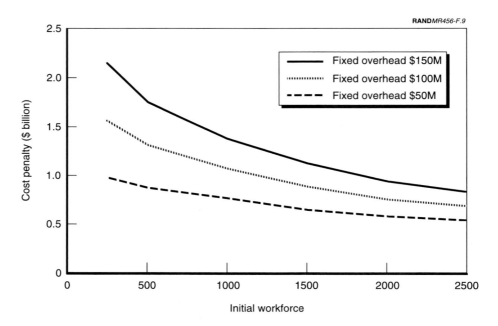

**Figure F.9—Workforce-Rebuilding Cost Penalty for a Sustained Production
Rate of Three Ships per Year**

figures show the cost penalty as a function of the size of the initial workforce, and present three cases for fixed annual overhead values of $50M, $100M and $150M.

The penalties range from roughly $150 million to more than $2 billion. The latter amount, for high rate and fixed overhead and low initial workforce, is equivalent to the cost of more than three and a half (steady-state) submarines. Across the ranges of parameter values tested, the penalty increases substantially with fixed overhead, even more so with production rate, and most of all with decreases in the initial workforce.

The results in Figures F.7 through F.9 were calculated with low early attrition rates. Increasing the early attrition from low to medium raises the cost penalty by approximately 10 percent (when initial workforce is low and fixed overhead is high; see Figure F.10). Going from medium to high results in a cost penalty increase of between 15 and 20 percent.

In contrast to the attrition rate, the mentor:trainee ratio has a considerable effect on the cost penalty. Raising the number of trainees per mentor from one to five cuts the cost penalty by more than half (for low initial workforce and early attrition and high fixed overhead; see Figure F.11). (The results shown above all assume a 1:2 ratio.)

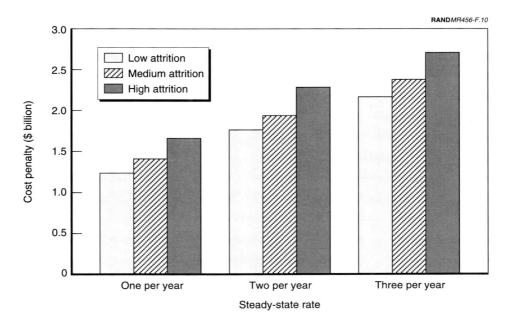

Figure F.10—Increasing Attrition Raises the Cost Penalty Slightly

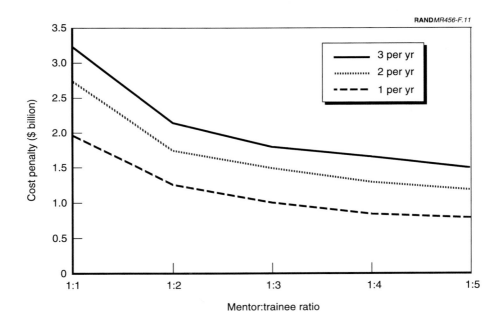

Figure F.11—The Cost Penalty Is Sensitive to the Mentor:Trainee Ratio

OPERATING AND SUPPORT COSTS

Appendix G presents background information regarding the operating and support (O&S) costs used in this study. The first section covers cost elements and definitions. The following three sections give specific values for the 688 class, Seawolf class, and NSSN, respectively.

ELEMENTS OF COST

Ship O&S costs are broken down into four major elements, with multiple sub-categories, in the Navy Visibility and Management of Operating and Support Costs (VAMOSC) system.[1] The following is a brief listing of the O&S elements:

Element

    1.0   Direct unit costs

        1.1   Personnel (officer and enlisted pay and allowances, and temporary additional duty costs)

        1.2   Material (petroleum, oil, and lubricants; repair parts; supplies; equipment/equipage; consumables; expendables; ammunition; and reparables)

        1.3   Purchased services

    2.0   Direct intermediate maintenance

        2.1   Afloat maintenance labor

        2.2   Ashore maintenance labor

---

[1]Naval Center for Cost Analysis, *Navy Visibility and Management of Operating and Support Costs (Navy VAMOSC), Individual Ships Report, Active Fleet Ships*, Volume 1, Report Description, DD-I&L (A&AR) 1422 (Ships 5200), June 1992.

2.3   Material

3.0   Direct depot maintenance

3.1   Scheduled ship overhaul (public and private shipyard costs for labor, material, and overhead to accomplish Regular overhauls (ROH), Engineered overhauls (EOH), Dry docking selected restricted availabilities (DSRA), and other overhaul categories).

3.2   Nonscheduled ship repairs (public and private shipyard costs for labor, material and overhead to accomplish restricted availabilities (RAV) and technical availabilities (TAV)).

3.3   Fleet modernization (public and private shipyard costs for labor, material and overhead to install ship alterations and improvements, including additional spares and other materials for the ship's coordinated shipboard allowance list).

3.4   Other depot (labor, material and overhead costs for naval aviation depot work; installation of modifications to equipment managed by the Space and Naval Warfare Systems Command; rework of ordnance equipment, hull, mechanical and electrical equipment, and electronic equipment; and design services allocation to cover drawing and technical data maintenance and updating).

4.0   Indirect operating and support

4.1   Training (cost of "C" and "F" course training for officers and enlisted).

4.2   Publications (replenishment of shipboard publications).

4.3   Engineering and technical services (cost of services provided outside of intermediate maintenance or depot availabilities).

4.4   Ammunition handling (cost of onload/offload by CONUS coastal handling stations).

VAMOSC does not currently include costs of two major depot activities for nuclear-powered vessels: refuelings, termed engineered refueling overhaul (ERO) for the 688 class, and decommissionings or inactivations (INAC).

Because this study is structured to consider alternative fleet sizes, construction schedules, and early retirement options, it is important to determine the cost and timing of particularly expensive depot maintenance activities. For nuclear attack submarines these include depot modernization periods (DMP), regular overhauls, refuelings, and inactivations. The remaining O&S costs were pooled to obtain an average annual O&S cost excluding the major depot availabilities. The following sections describe how these values were determined for this study.

688-CLASS O&S COSTS

Table G.1 summarizes the O&S costs and timing for submarines in the 688 class. In terms of depot activities and schedules, there are three subsets of 688-class submarines, which are indicated by the hull number groupings: 688–699, 700–718, and 719–773. The scheduled times for major depot availabilities were taken from OPNAVNOTE 4700, dated December 2, 1992. Cost numbers in the table were developed from VAMOSC data for 688-class submarines (average annual O&S, overhauls, and depot modernization) or provided by the Naval Nuclear Propulsion Program (refueling and inactivation). Drydocking selected restricted availabilities have not been treated separately because they are of relatively short duration and low cost. Their costs are included in the average annual O&S cost.

Table G.1

688-Class Submarine O&S Costs (FY91$)

Description	Cost ($ million)	Hull Numbers	Timing
Average annual O&S Cost (excluding major depot availabilities)	$15	All	Annual
Regular overhaul (ROH)	$175	688–699	7th year
Depot modernization period (DMP)	$90	700–773	7th year
Engineered refueling overhaul (ERO)	$265	688–718	16th year
		719–773	24th year
Engineered overhaul (EOH)	$175	688–718	24th year
		719–773	16th year
Inactivation (INAC)	$50	All	30th year

Complete inactivation includes three major steps. The first, the inactivation availability, includes reactor defueling; systems shutdown; removal of equipment that can be used by the active fleet; and preparation for waterborne storage if recycling will not take place immediately, modifications to permit towing to Puget Sound Naval Shipyard (if not there already), and missile compartment dismantlement for ballistic missile submarines. The second is removal and disposal of the defueled reactor compartment. The third, recycling, involves total dismantlement of the remaining portions of the submarine, with useful equipment put into inventories for possible future use and everything else sold for scrap. The $50M represents a typical value covering all three steps for all 688-class submarines.[2]

SEAWOLF-CLASS O&S COSTS

The Seawolf-class submarines will have a propulsion system that will last the life of the ship (30 years) without refueling. For the analyses in this study, the Seawolf-class O&S costs are derived from the O&S section in the December 1992 Selected Acquisition Report (SAR). The annual depot cost was multiplied by 30 to obtain the total lifetime depot cost. This was then divided by three to obtain a representative overhaul cost to apply in the 7th, 16th, and 24th years, following the pattern of the 688 class.[3] This yielded $200M per overhaul, in round numbers. The remaining costs ($17M per year) were used as the annual O&S cost. It was assumed that Seawolf inactivation would not be significantly different from the 688 class and $50M was used as the inactivation cost.

NSSN O&S COSTS

The NSSN is undergoing initial concept definition and its characteristics are not yet fully identified. Alternatives being considered range from roughly the 688 class to the Seawolf class. For this study, it was assumed that the NSSNs would have a power plant that would last the life of the submarine. All costs were assumed to be similar to the 688 class: average annual O&S of $15M, overhaul costs of $175M (occurring at the 7th, 16th, and 24th years[4]), and inactivation costs of $50M.

---

[2]This value is based on the planning estimate for inactivation of SSN-689 at Mare Island Naval Shipyard plus completion of the total inactivation at Puget Sound. It does not reflect taking advantage of the economies suggested in the General Accounting Office (GAO) report *Nuclear Submarines: Navy Efforts to Reduce Inactivation Costs*, GAO/NSIAD-92-134, July 1992.

[3]The Navy has recently informed us that there will be only one overhaul for the Seawolf class, at midlife. The effect of this change on our cost analysis would be negligible, as the total depot cost would be the same.

[4]The previous note applies also to the NSSN.

COMPARING COSTS: ADDITIONAL CASES

Chapter Seven examines cost estimates for alternative production gap and ship-life strategies intended to maintain a fleet of 40 submarines at a maximum production rate of two ships per year. In Appendix H, we present analogous graphs (with some commentary) for strategies that could maintain fleet sizes of 40 or 50 ships at maximum production rates of three per year. As in Chapter Seven, we show cumulative costs to 2030, discounted and undiscounted, in absolute terms and relative to the minimum-gap, 30-year-life case, and bar charts that break down the undiscounted-relative-cost charts into categories (construction, refueling, reconstitution, and overhauls).

SUSTAINING A FLEET SIZE OF 40 SHIPS AT THREE PRODUCED PER YEAR

Figures H.1 through H.4 show the cumulative costs of sustaining the submarine fleet (688s, Seawolves, and NSSNs) from 1994 through 2030. The costs depicted in these figures follow the general pattern in the analogous two-per-year graphs (Figures 7.2, 7.4, 7.6, and 7.7), with the exception of the 30-year-ship-life, maximum-gap strategy. At two ships per year, the latter differs little from the corresponding minimum-gap case, because at such a low production rate, a gap cannot be long if the force is not to drop below 40 ships. If it is possible to build three ships per year after restart, a longer gap can be opened up. Thus, in the current comparisons, the 30-year-life, maximum-gap case shows a cost pattern close to the other strategies (maximum gap or 35-year life) entailing postponed production. That is, savings accrue from a construction rate that is initially below that in the baseline case. When high-rate construction begins under the postponed-production strategies, some or all of that advantage is lost (less in discounted terms than in undiscounted terms).

The general lesson is the same as for the two-per-year case. Extended gaps in production do not result in large savings. For the current 30-year-life, there is essentially no difference between the minimum and maximum gaps over the

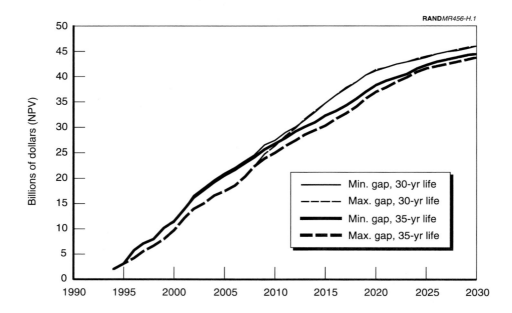

Figure H.1—Cumulative Cost of Sustaining the Attack Sub Fleet
(40-Ship Minimum) at a Maximum Production Rate of Three
per Year, Discounted at 5 Percent per Year

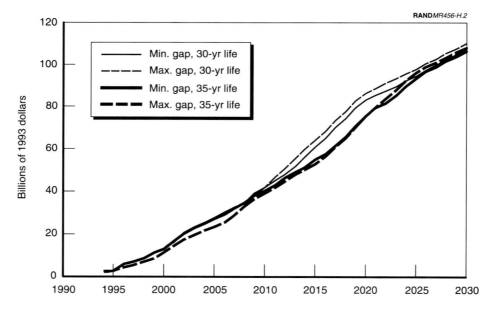

Figure H.2—Cumulative Cost of Sustaining the Attack Sub Fleet
(40-Ship Minimum) at a Maximum Production Rate of
Three per Year, Undiscounted

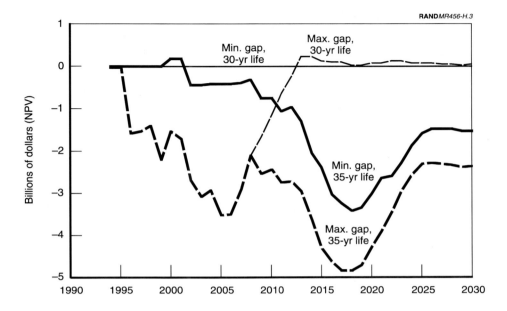

Figure H.3—Cumulative Cost of 40-Ship, Three-per-Year Strategies, Relative to Cumulative Cost for Min.-Gap, 30-Year-Life Strategy, Discounted

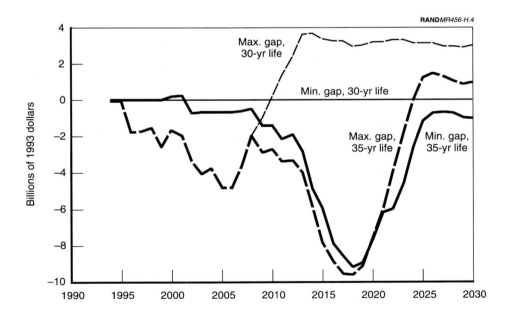

Figure H.4—Cumulative Cost of 40-Ship, Three-per-Year Strategies, Relative to Cumulative Cost for Min.-Gap, 30-Year-Life Strategy, Undiscounted

long term; with life extension, the longer gap saves roughly a billion dollars. For equivalent gapping strategies, life extension saves on the order of $2 billion.

Figures H.5 through H.8 are the analogues of Figures 7.8, 7.10, 7.11, and 7.12 for the two-per-year case and are useful in examining the sources of the cost differences. The sharpest differences between the two- and three-per-year cases are in the two pairs of graphs (first and last) involving the 30-year, minimum gap strategies. The anomalous nature of that strategy for the two-per-year case has already been noted. However, some differences are apparent in all comparisons because the ability to build three ships per year permits greater postponement of production in the maximum-gap and 35-year strategies, resulting in wider swings in costs.

SUSTAINING A FLEET SIZE OF 50 SHIPS AT THREE PRODUCED PER YEAR

Costs of the 50-ship, three-per-year strategies are compared in Figures H.9 through H.12. To repeat here for the reader's convenience the points made in Chapter Seven, the total cost of sustaining a 50-ship fleet is, of course, more than that of sustaining a 40-ship fleet. However, the relations among the strategies are similar in the two cases (compare Figure H.9, for example, with

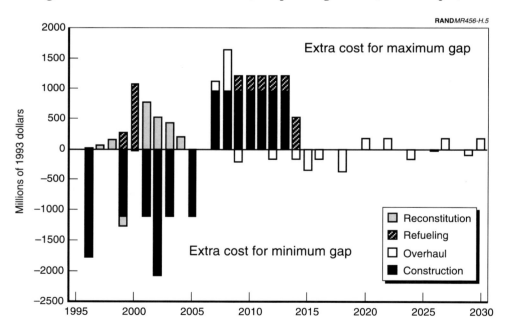

Figure H.5—Annual Cost Difference, Min. vs. Max. Gap, Max. Ship Life 30 Years, Fleet Size 40, Max. Rate Three per Year

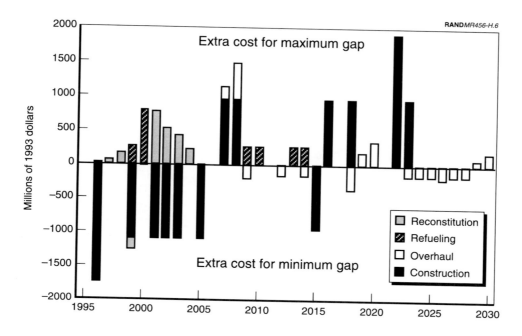

Figure H.6—Annual Cost Difference, Min. vs. Max. Gap, Max. Ship Life 35 Years, Min.
Fleet Size 40, Max. Production Rate Three per Year

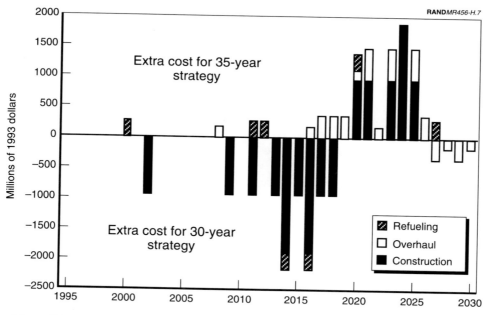

Figure H.7—Annual Cost Difference, 30-Year Max. Ship Life vs. 35 Years, Min. Gap,
Min. Fleet Size 40, Max. Production Rate Three per Year

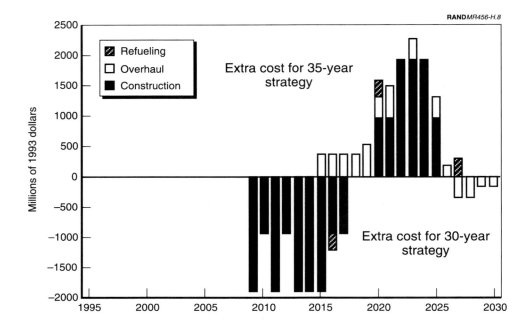

Figure H.8—Annual Cost Difference, 30-Year Life vs. 35 Years, Max. Gap, Min. Fleet
Size 40, Max. Production Rate Three per Year

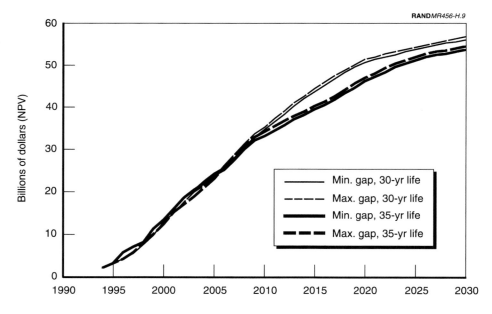

Figure H.9—Cumulative Cost of Sustaining the Attack Submarine Fleet
(50-Ship Minimum) at a Maximum Production Rate of Three per Year,
Discounted at 5 Percent per Year

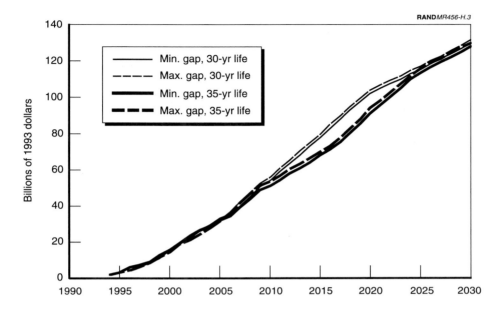

Figure H.10—Cumulative Cost of Sustaining the Attack Submarine Fleet (50-Ship Minimum) at a Maximum Production Rate of Three per Year, Undiscounted

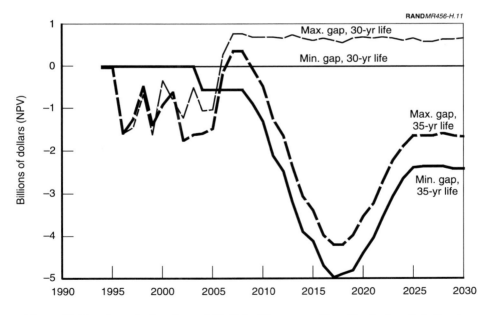

Figure H.11—Cumulative Cost of 50-Ship, Three-per-Year Strategies, Relative to Cumulative Cost for Minimum-Gap, 30-Year-Life Strategy, Discounted

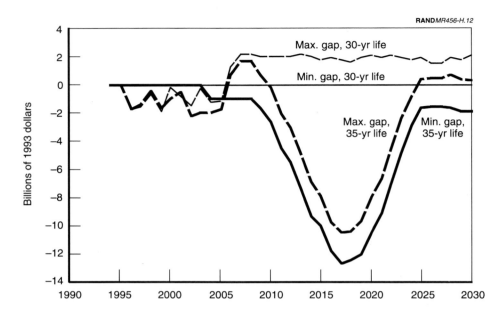

Figure H.12—Cumulative Cost of 50-Ship, Three-per-Year Strategies, Relative to
Cumulative Cost for Minimum-Gap, 30-Year-Life Strategy, Undiscounted

Figure H.1). Again, the costs of the strategies are all within a few billion dollars
of each other, and changing maximum ship life has a bigger effect than changing the length of the production gap.

The biggest difference from the 40-ship case is that the maximum-gap strategies are no longer at parity or at an advantage with respect to the minimum
gaps. The differences are not large, but they are consistent. For the larger fleet
size, the maximum gap strategy with life extension is about a billion-and-a-half
dollars worse off with respect to the corresponding minimum-gap strategy. The
small advantage for 40 ships becomes a small disadvantage for 50. For the 30-
year case, the maximum-gap strategy is about $700 million worse off. Larger
costs are run up relative to the minimum gap because production cannot be
put off as long when a 50-ship fleet must be replaced as it can when only 40
need be built.

Because production gaps do not have the effects on schedule when a 50-ship
fleet must be sustained that they do when only 40 ships are needed, the min.-
max. comparisons differ greatly between the two cases (compare Figure H.13
with Figure H.5 and Figure H.14 with Figure H.6). The inability to concentrate
production in the maximum-gap strategy is clearly visible in Figures H.13 and
H.14. This does not, however, apply to the 30-vs.-35-year comparisons. At 50

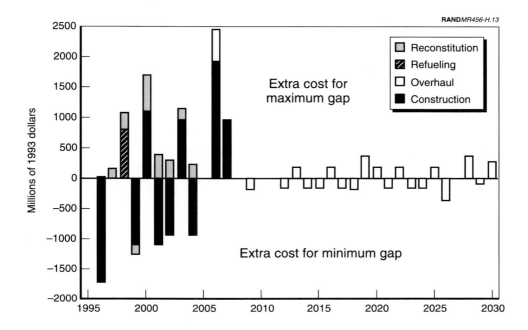

Figure H.13—Annual Cost Difference, Minimum vs. Maximum Gap, Max. Ship Life 30 Years, Min. Fleet Size 50, Max. Production Rate Three per Year

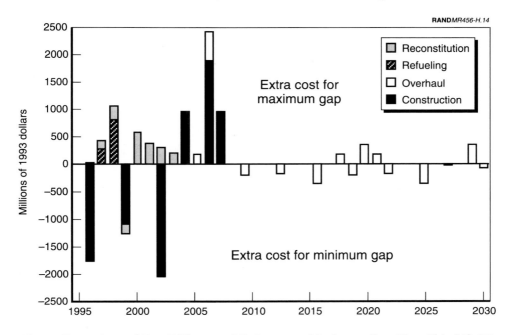

Figure H.14—Annual Cost Difference, Minimum vs. Maximum Gap, Max. Ship Life 35 Years, Min. Fleet Size 50, Max. Production Rate Three per Year

ships, those comparisons show the same postponement of production as was achievable with 50 ships. The 30-vs.-35 cost swings are evident both in Figure H.12 and in Figures H.15 and H.16.

The reason for the difference is that the extra five years allow more "breathing room." In the 35-year strategies, production can fall to one a year between about 2010 and 2016, when the force size can be maintained by refueling ships hitting their 24-year mark. No such lull is possible in the 30-year strategies, under which all the replacement ships need to be started by 2021. The lull in the 35-year strategies allows a money-saving postponement of production into the 2020s.

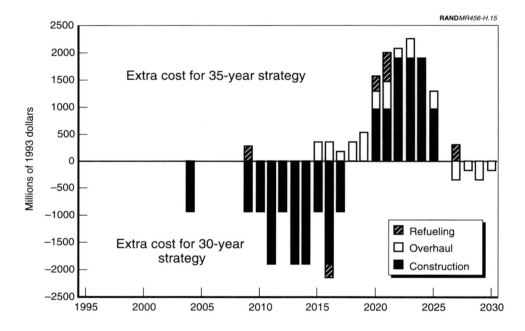

Figure H.15—Annual Cost Difference, 30-Year vs. 35-Year Max. Ship Life, Min. Gap, Min. Fleet Size 50, Max. Production Rate Three per Year

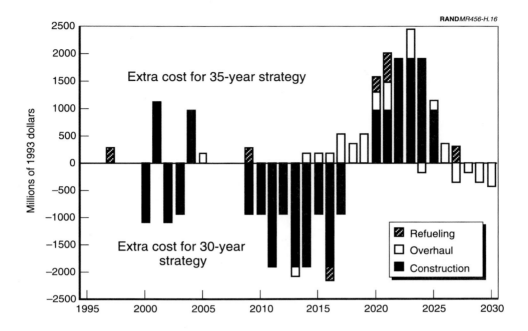

Figure H.16—Annual Cost Difference, 30-Year vs. 35-Year Max. Ship Life,
Max. Gap, Min. Fleet Size 50, Max. Production Rate Three per Year